THE SENIOR ISSUES COLLECTION

COLLECTION EDITOR:
GILLDA LEITENBERG

FAMILY ISSUES

EDITED BY KATHY EVANS AND
GILLDA LEITENBERG

To Dale, Jordan, and Joel,
for letting me be a part of them.
Kathy

To Sam: my family, my all.
Gillda

McGraw-Hill Ryerson Limited

Toronto • Montreal • New York • Auckland • Bogotá
Caracas • Lisbon • London • Madrid • Mexico • Milan
New Delhi • San Juan • Singapore • Sydney • Tokyo

Family Issues
The Senior Issues Collection

Copyright © McGraw-Hill Ryerson Limited, 1995
All rights reserved. No part of this publication may be reproduced or transmitted in any
form or by any means, or stored in a data base or retrieval system, without the prior
written permission of McGraw-Hill Ryerson Limited.

ISBN 0-07-551696-9

2 3 4 5 6 7 8 9 10 BG 04 03

Printed and bound in Canada

Canadian Cataloguing in Publication Data

Main entry under title:

Family issues

(The Senior issues collection)
ISBN 0-07-551696-9

1. Readers (Secondary). 2. Readers – Family.
3. Family – Literary collections. I. Evans, Kathy,
date. II. Leitenberg, Gillda. III. Series.

PE1127.F35F35 1995   808'.0427   C95-930498-3

Editor: *Kathy Evans*
Supervising Editor: *Nancy Christoffer*
Permissions Editor: *Jacqueline Donovan*
Copy Editor: *Gail Marsden*
Designer: *Mary Opper*
Typesetter: *Pages Design Ltd.*
Photo Researcher: *Elaine Freedman*
Cover Illustrator: *Rocco Baviera*

The editors wish to thank reviewer Freda Appleyard for her comments and advice.

This book was manufactured in Canada using acid-free and recycled paper.

Contents

Introduction

Robert Frost, in writing about home, called it, "Something you some-how haven't to deserve." You may live with both parents, or a single parent, or grandparents, or foster parents. You may have a step-parent. Or you may have lost touch with blood relatives, and your friends or your community may have become your home. No matter the configuration or the situation, we all have a family—we "haven't to deserve one"—and that's what this book is about.

As you read, you'll find many family portraits: a woman mourning the loss of her mother; a father teaching a daughter to drive; a grand-father acting as a single parent; a community rescuing a senior citizen who has lost her home in a fire. You'll meet an adopted boy who has questions about his past; a woman who has helped to raise her disabled sister; a woman who remembers the tension between her parents at the family dinner table; a man who worries about his sons' fighting; and men and women who question gender roles in families.

Many of the families are facing changes: divorce, aging, illness, and clashes in intergenerational values and cultures. Other families are facing the realities of war, poverty, racism, and abuse. As in families themselves, the anthology contains funny moments, times of happiness, and times of extreme sorrow.

Currently there's a sense that families are dissolving and breaking apart, and that family values are disintegrating. The media, especially television, certainly depict many families facing transition. As you explore the issues raised in the readings, you will shape your own opinions about the state of the family. Independent study and cross-curricular inquiries will lead you to explore some of the issues in greater depth and from new perspectives.

You'll notice that the cover of this anthology reflects many of the issues addressed in the contents: culture, roots, generations, relation-ships, change, and, at the centre, self. Keep yourself there, in the centre, as you read. You might make discoveries about your own circumstances. You'll also consider other families and be sensitive to global family concerns.

There are many changes that can affect family relationships, but what will sustain us—and what is lasting—is love and respect for one another.

Enjoy.

Kathy Evans and Gillda Leitenberg

Offered and Taken

BY

DOMINIQUE
LABAW

I took my father up in a dream
Above the planet to look down upon his own reflection
In my eyes
And in those of my brothers, who came too
And in the quiet faces of my sisters, who flew to meet us
The years we spent not seeing it
Our sameness
How it moves below the dermis
Owning all the hereditary maybes
Pulling the many out into the open
High and higher over us
Beyond anything corporeal
Whispers of new ones, of Allah, of the yet-to-be
Growing right here in our hands
And we are Isaac and Frances and Lurlene and Dawud and
    Dominique and
Ishaq and Jalil and Ayasha and Yasmin and Lisa and Ma Aloom
    Jama
And yes, there are others
Each one a lightyear, a living sun
A star.

What Happened to Family?

BY

LESLEY

FRANCIS

~

After exchanging their wedding vows, Nancy and Rick Davidson took out two tiny rings. Mom and Dad gave the rings to the two little girls in their wedding party—Amanda, 6, who was the flower girl, and Leah, 3, the ring bearer. The two adults and the two children (Nancy's daughters from a previous marriage) were all starting a new life.

"It was our way of saying we're a family now," says Nancy.

Since that day seven years ago, the Davidsons have been doing what all Canadian families do, no matter what form they take—strengthening the bond they have with each other. Their blended family is just one example of the different kinds of families Canadians live in today. That diversity in the 1990s can be confusing and stressful for people who think all families should be like the ideal family of the 1950s.

"It's far more important to look at what families do than what they look like," says Robert Glossop, director of programs and research for the Vanier Institute of the Family in Ottawa. "Families are about shared biographies, about lives lived together, about the experience of one becoming the experience of others."

However, many of us can't shake the 1950s ideal, says family sociologist Dr. Barrie Robinson in an interview from Vancouver. And we react negatively to the diversity of family forms we see around us.

"However, there is no eternal model of family that transcends time and place."

In fact, there have always been a variety of families, says Glossop. We may be faced with a lot of change in the 1990s but there is also lots of continuity.

For instance, single-parent families aren't new. In most cultures at most times one in three women did not have a partner.

Sixty years ago, 13 per cent of families had single parents. Today the figure is the same. But during the 1950s that number dropped to between eight and nine per cent, says Glossop.

And women have always contributed to the family income, whether they lived on farms or in industrial areas. The 1950s represent the only period of time when that wasn't true for most families.

Nor is the nuclear family an invention of the 20th century. It could be found in 14th century Italy or 16th century France as long as it's defined as the basic family unit—adults and dependent children.

"What changes is the context it exists in," Glossop says. "It might have been part of an extended kinship network with other members of family living next door to you. In native Canadian cultures, it's hard to tell where the family ends and the community begins."

The nuclear family of the 1950s was cut off from relatives and other supports, says Glossop. It was expected to be mobile and ready to go at a moment's notice.

Today's nuclear family remains in that isolation. It is expected to be self-sufficient and self-reliant. Combine this with the changes in the size of family, wage-earning patterns, rates of divorce and separation, conflicting expectations of what can be accomplished in raising kids, and the push to be more material. What you end up with is a picture of the family struggling to survive, says Glossop.

"Families are not closed systems. They aren't immune to the pressures of the outside world. You have to be a Supermom, a Superdad or a Superkid. Parents have to teach their kids to be competitive but also caring, compassionate and willing to share."

The family has become the only place people expect to find love in our society, says American social historian Stephanie Coontz from Olympia, Wash. Glossop acknowledges that it has always been tough for the family. And trying to ease things for the family of the 1990s is about as tough as it gets.

"We need to become a more familial society," he says. "Expand our notion of connectedness beyond our immediate family to include friends, neighbors, etc. rather than adopt this proprietary notion about our kids."

"The more we pretend one form is perfect while others are terrible only exacerbates the problems, pressures and stresses for all families," Coontz says. "We end up in a feeding frenzy of guilt and blame."

Friends and Family

~

BY

SUSAN

ROGERS

The phone rings and you're asked to pull together 30 family members and friends for a photograph—quickly. Who would be included in this photo of your extended family?

Such a request by *Canadian Living* was a snap for Debbie and Mike Campbell of Belmont, N.S., who have enough parents, sisters, brothers, nieces and nephews to fill any wide-angle lens. Although some of the 13 siblings they have between them live elsewhere, the Campbells easily assembled more than 30 family members and a few friends with only three days'

notice. Soccer games were cancelled and dinner was postponed, but they gladly gathered in Halifax at Debbie's bidding.

"Just seeing each other is a joy," says Debbie, accustomed to frequent family gatherings, where everyone brings food and the kids have a ball. "My extended family is very important to me. We've always had one another."

But many families today are neither as large nor as close, in both senses of the word. To organize a large family photo, many of us would have to issue several airline tickets. Or we could invite lots of friends because, for many of us, friends are family.

The average Canadian household has 2.7 members, according to the 1991 census. What few kin we may have are often scattered across this vast nation or have been left behind in other countries. Family ties are stretched by long distances. Almost one million Canadians over the age of 50 do not have frequent contact with close relatives.

Nevertheless, kinship continues to be important in the day-to-day life of the average Canadian, concludes Thomas Burch, a sociologist at the University of Western Ontario in London, Ont. After studying data from a 1990 survey of national social trends, he found most

Family and friends of Debbie and Mike Campbell

Canadians have at least six close relatives throughout their lives. A majority live close to some of their kin and see them frequently. Based on other research, he estimates 30 to 50 per cent of urban Canadian families see a brother or sister, aunt or uncle once a week.

Divorce has added yet another dimension to the concept of extended family. When divorced parents remarry, all kinds of new relatives are grafted to the family tree. "Some children have more grandparents and siblings than ever before," says Burch. But would they want to be in the same room, let alone the same family photo?

Sociologist Barry Wellman says we tend to glorify the extended families of the old days. "People have as much kinship now," he claims. "They're not having as many children but they are staying alive longer. We can be in touch with family by phone, plane, car and other ways they couldn't 100 years ago."

Wellman's research at the University of Toronto's Centre for Urban and Community Studies shows a strong relationship between mothers and adult daughters and between brothers and sisters. "But there's hardly anything with aunts and uncles," he says.

We'll Impose on Family

Ann Phelan of East River, N.S., is the exception to such sociological findings. Married with no children, Phelan nurtures a strong bond with a bevy of "aunties" in her life. Her mother, who died when she was 16, had 11 siblings and, though estranged from her father, Phelan is close to his sisters, too. "Because I had their unwavering support and love as a child, I have it for them now," says Phelan, now 40. "I will respond to them immediately. It's automatic." Not that Phelan doesn't struggle with "the burden of guilt" when dealing with aging family members who sometimes don't recognize the demands of her busy life. "Their need is louder," she says. "It's hard to say no."

On the flip side, Phelan and her husband, Bruce, feel comfortable asking for help from their cousins. Such ready assistance was invaluable while they were building their house on the ocean. "We could have called friends but we didn't want to impose," says Phelan. "It was easier to call a cousin who would say 'Don't worry,' and six cousins would show up on the weekend."

This tendency to rely on family rather than friends is typical. Most of us are more willing to impose on family members, if given the choice. Psychologist Brian de Vries of the University of British Columbia in Vancouver says people want a reciprocal relationship with friends. Once it becomes unbalanced, they fall back on family.

Phelan also plays a strong role in the lives of two of her godchildren and has good relationships with another two, although she doesn't see them as often. Casting herself as "an alternative role model," Phelan spends "quality time" with these young people in her life. They may spend a weekend horseback riding together or simply a couple of hours strolling along the beach looking for seashells. The goal is to create opportunities when the youngsters can talk and "trust unequivocally." A couple of the children call her Auntie Ann although they aren't connected by blood.

Described as "fictive kin" in sociological circles, such adopted aunts and uncles obscure the distinctions between friends and family. Studies have revealed that people will include friends—and even pets—as family, says de Vries. "There is a blurring of friends and family because large numbers identify friends as family and family as friends."

According to a 1990 survey of Canadian social trends, friends now take on a bigger role in our lives because the average Canadian has seven or eight close friends and only four siblings. While family tends to help most with money and housework, friends pitch in, too. In fact, chances are a friend is more likely to give you a ride or help with household maintenance.

Building Traditions

No one knows the importance of friends more than Sheila Costello, who emigrated from Ireland 22 years ago, leaving behind her parents, four brothers and four sisters. Within a year, she developed a network of friends in Toronto, mostly among other immigrants. They remain "fiercely loyal to this day," says Costello. "We would view one another as family, warts and all."

This "family" has expanded to include spouses and children who, along with the "never-marrieds," continue to gather for holidays. "We have built our own traditions," says Costello, who insists on sitdown meals and pageantry for Thanksgiving and Christmas. Ironically, her family in Ireland doesn't celebrate such occasions together. But now that her sister Mary lives in Nova Scotia, the two do call and get together as often as the 1860-km separation will permit.

"In many ways you can choose your friends and not your family," Costello says honestly. "But the same pitfalls are there with friends. You can take each other for granted. And you don't always like them. We need to work on friendships."

Lorraine Smith-Collins, on the other hand, says, "Family are our friends." With four children of their own and close ties to extended family, she and her husband, Butch, have little time to socialize. As soon as their first child was expected, they moved back to Hantsport, N.S., to be near her parents and five of the eight siblings still living in the area. "We value having the children close to the grandparents," she says. "And we have a great need for them to know their cousins."

Part of that need is cultural because Lorraine is Micmac. Her parents, Rita and Noel Smith, now live alone in the family's white clapboard home on the small Horton Reserve, created in a political split from another band by Rita, who was also chief for three years. There they are surrounded by family. Photographs of their nine children, 27 grandchildren and sundry great-grandchildren adorn the walls of two entire rooms. A daughter and son live in two bungalows across the road.

Family contact isn't all celebrations and story-telling sessions, however. These days, meetings are held to ensure that Rita and Noel, now suffering from medical problems, receive the care they need to remain in their home. Brothers and sisters take turns preparing meals for their parents.

Lorraine is rewarded when an 18-year-old nephew comments on how lucky they all are. "He said a lot of his friends don't have such close families," she says. "I think it's wonderful he recognizes the connections."

Kinship Found in Community

For Canadians without those connections, kinship can sometimes be found in community. Resource centres and neighbourhood groups provide their own versions of family.

Apple Tree Landing Children's Centre, one of many family resource centres emerging across the country, certainly extends itself to residents around Canning, a small community in Nova Scotia's Annapolis Valley. The four MacIntosh red awnings over the large windows in the centre's mainstreet building are as welcoming as, well, Mother's apple pie. Besides providing a toy and book library for children, nursery school and after-school care, Apple Tree Landing is a place where parents and caregivers are able to congregate.

"I do think such a centre is about trying to create links where there aren't links," says president of the board and founding member Carole MacInnes, originally from Arkansas. She sees these links spreading into the community, a community with a lot of come-from-aways like herself. Families create networks of friends, the kind "who offer help when you need it before you have to ask."

For the million single parents in Canada, such institutions often provide badly needed child care and support. Although relatives still look after 40 per cent of the children under 13 in all families, single moms can find themselves very much alone.

Parent resource centres, such as the one in Dartmouth, N.S., are places where single parents can seek advice and companionship. "It's like calling Mom," says coordinator Linda Wentzell, who has been known to spend six hours on New Year's Eve with a distraught mother. "People who come here often have no family around. They say, 'This is my family.'"

Cathy Seale of Dartmouth does have family and lots of it in Cape Breton. But when she separated from her husband for the second time, she decided she couldn't go home again with her two sons. "It was hard at first staying in Dartmouth as a single mother away from my family," she says. "Things come up that you would like your family around for."

Parents, siblings and in-laws still give financial and moral support to Seale and her sons. Day-to-day help, however, comes from the moms she has met at the Dartmouth centre. If she arranged a family photo, it would have to include these people, who are always ready to lend a loaf of bread or a sympathetic ear. "The resource centre is my family now," says Seale. "I'd be lost without the women here."

Mothers

BY

NIKKI

GIOVANNI

the last time i was home
to see my mother we kissed
exchanged pleasantries
and unpleasantries pulled a warm
comforting silence around
us and read separate books

i remember the first time
i consciously saw her
we were living in a three room
apartment on burns avenue

mommy always sat in the dark
i don't know how she knew but she did

that night i stumbled into the kitchen
maybe because i've always been
a night person or perhaps because i had wet
the bed
she was sitting on a chair
the room was bathed in moonlight diffused through
those thousands of panes landlords who rented
to people with children were prone to put in windows

she may have been smoking but maybe not
her hair was three-quarters her height
which made me a strong believer in the samson myth
and very black

she was very deliberately waiting
perhaps for my father to come home
from his night job or maybe for a dream
that had promised to come by
"come here" she said "i'll teach you
a poem: *i see the moon*
            the moon sees me
            god bless the moon
            and god bless me"
i taught it to my son
who recited it for her
just to say we must learn
to bear the pleasures
as we have borne the pains

Whose Mouth Do I Speak With

~

BY

SUZANNE

RANCOURT

I can remember my father bringing home spruce gum.
He worked in the woods and filled his pockets
with golden chunks of pitch.
For his children
he provided this special sacrament
and we'd gather at his feet, around his legs,
bumping his lunchbox, and his empty thermos rattled inside.
Our skin would stick to Daddy's gluey clothing
and we'd smell like Mumma's Pine Sol.
We had no money for storebought gum
but that's all right.
The spruce gum
was so close to chewing amber
as though in our mouths we held the eyes of Coyote
and how many other children had fathers
that placed on their innocent, anxious tongues
the blood of trees?

from

generations

~

BY

LUCILLE

CLIFTON

Oh she made magic, she was a magic woman, my Mama. She was not wise in the world but she had magic wisdom. She was twenty-one years old when she got married but she had had to stay home and help take care of her brothers and sisters. And she had married Daddy right out of her mother's house. Just stayed home, then married Daddy who had been her friend Edna Bell's husband after Edna Bell died. She never went out much. She used to sit and hum in this chair by the window. After my brother was born, she never slept with my Daddy again. She never slept with anybody, for twenty years. She used to tell me "Get away, get away. I have not had a normal life. I want you to have a natural life. I want you to get away."

A lot of people were always telling me to get away.

She used to sit in this chair by the window and hum and rock. Some Sundays in the summertime me and her used to go for walks over to the white folks section to look in their windows and I would tell her when I grew up I was going to take her to a new place and buy her all those things.

Once in a while we would go to the movies, me and her. But after she started having her fits I would worry her so much with Are you all right Ma and How do you feel Ma that we didn't go as often. Once I asked her if she was all right and she said she would be fine if I would leave her alone.

Mostly on Friday nights when Daddy had gone out and the other kids had gone out too we would get hamburgers and pop from the store and sit together and after we got TV we would watch TV. On New Year's Eve we would wait up until midnight and I would play Auld Lang Syne on the piano while me and my Mama sang and then we would go to bed.

Oh she was magic. If there were locks that were locked tight, she could get a little thing and open them. She could take old bent hangers and rags and make curtains and hang drapes. She ironed on chairs and made cakes every week and everybody loved her. Everybody.

When Daddy bought the house away from Purdy Street, Mama didn't know that he had been saving his money. One day he just took us to see this house he was buying. I was going away to college that fall and Punkin was off and married and we were scattering but he had bought us this house to be together in. Because we were his family and he loved us and wanted us to be together. He was a strong man, a strong family man, my Daddy. So many people knew him for a man in a time when it wasn't so common. And he lived with us, our Daddy lived in our house with us, and that wasn't common then either. He was not a common man. Now, he did some things, he did some things, but he always loved his family.

He hurt us all a lot and we hurt him a lot, the way people who love each other do, you know. I probably am better off than any of us, better off in my mind, you know, and I credit Fred for that. Punkin she has a hard time living in the world and so does my brother and Jo has a hard time and gives one too. And a lot of all that is his fault, has something to do with him.

And Mama, Mama's life was—seemed like—the biggest waste in the world to me, but now I don't know, I'm not sure any more. She married him when she was a young twenty-one and he was the only man she ever knew and he was the only man she ever loved and how she loved him! She adored him. He'd stay out all night and in the morning when he came home he'd be swinging down the street and she would look out the window and she'd say loud "Here comes your crazy Daddy." And the relief and joy would make her face shine. She used to get up at five every morning to fix his breakfast for him and she one time fell down the back steps and broke her ankle and didn't see about it until after she had fixed breakfast, had gotten back up the steps and finished.

She would leave him. She would leave him and come in every morning at five o'clock to fix his breakfast because "your Daddy works hard," she would fuss, "you know you can't fix him a decent meal."

She would sit in the movies. She would leave him and sit in the movies and I would see her there and try to talk and make things right. I always felt that I was supposed to make things right, only I didn't know how, I didn't know how. I used to laugh and laugh at the dinner table till they thought I was crazy but I was so anxious to make things right.

I never knew what to do. One time they were arguing about something and he was going to hit her and my sister Punkin, who had a different mother, she ran and got the broom and kept shouting "If you hit Mama I'll kill you" at Daddy. My brother and I didn't do anything but stand there and it was our Mama but we didn't do anything because we didn't know what to do.

Another time they were arguing and I was in the kitchen washing dishes and all of a sudden I heard my Mama start screaming and fall down on the floor and I ran into the room and she was rolling on the floor and Daddy hadn't touched her, she had just started screaming and rolling on the floor. "What have you done to her," I hollered. Then "What should I do, what should I do?" And Daddy said "I don't know, I don't know, I don't know, she's crazy," and went out. When he left, Mama lay still, and then sat up and leaned on me and whispered "Lue, I'm just tired, I'm just tired."

The last time ever I saw her alive she had been undergoing tests to find out what caused her epilepsy and I leaned over to kiss her and she looked at me and said "The doctors took a test and they say I'm not crazy. Tell your Daddy."

I wanted to make things better. I used to lay in bed at night and listen for her fits. And earlier than that, when I was younger, a little girl, I would lay awake and listen for their fights. One night they were shouting at each other and my sister Punkin whispered out of her stillness "Lue, are you awake?" "No," I mumbled. She stirred a little. "That's good," she said.

I wanted to make things right. I always thought I was supposed to. As if there was a right. As if I knew what right was. As if I knew.

My Mama dropped dead in a hospital hall one month before my

first child was born. She had gone to take a series of tests to try to find out the cause of her epilepsy. I went to visit her every day and we laughed and talked about the baby coming. Her first grandchild. On this day, Friday, February 13, it was raining but I started out early because I had not gone to see her the day before. My aunt and my Uncle Buddy were standing in the reception area and as I came in they rushed to me saying "Wait, Lue, wait, it's not visiting hours yet." After a few minutes I noticed other people going on toward the wards and I started up when my aunt said "Where are you going, Lue?" and I said "Up to see my Mama," and they said all together "Lue Lue your Mama's dead." I stopped. I said "That's not funny." Nobody laughed, just looked at me, and I fell, big as a house with my baby, back into the telephone booth, crying "Oh Buddy Oh Buddy, Buddy, Buddy."

~

One month and ten days later another Dahomey woman was born, but this one was mixed with magic.

Things don't fall apart. Things hold. Lines connect in thin ways that last and last and lives become generations made out of pictures and words just kept. "We come out of it better than they did, Lue," my Daddy said, and I watch my six children and know we did. They walk with confidence through the world, free sons and daughters of free folk, for my Mama told me that slavery was a temporary thing, mostly we was free and she was right. And she smiled when she said it and Daddy smiled too and saw that my sons are as strong as my daughters and it had been made right.

And I could tell you about things we been through, some awful ones, some wonderful, but I know that the things that make us are more than that, our lives are more than the days in them, our lives are our line and we go on. I type that and I swear I can see Ca'line standing in the green of Virginia, in the green of Afrika, and I swear she makes no sound but she nods her head and smiles.

> *The generations of Caroline Donald born in Afrika in 1823*
> *and Sam Louis Sale born in America in 1777 are*
> *Lucille*
> *who had a son named*
> *Genie*
> *who had a son named*

Samuel
who married
Thelma Moore and the blood became Magic and their
daughter is
Thelma Lucille
who married Fred Clifton and the blood became whole and
their children are
Sidney
Fredrica
Gillian
Alexia four daughters and
Channing
Graham two sons,
and the line goes on.
"Don't you worry, mister, don't you worry."

Father's Gloves

BY

PAUL

WILSON

I didn't expect this warmth.
As if they jostled in the box,
vacant thumbs to vacant fingers,
palms rubbing counter to the moon,
gloves forming a season of their own.

None match. A few seem too small
for your hands, or mine. A finger nail
has worried fabric till threads snapped. Here work
moved out from your body, the arm of the jack recoiled,
bruising the tender heel of your hand.

These gloves fold in on themselves. Leather fingers
clung to your skin, reversing as you withdrew to stroke
a child's hair, a vein on a poplar leaf. I see
your hand, tense on an axe handle, a bucksaw.
With this glove you wiped grease from your wrist.

Your gloves carry lines of your life: the burn of
baling twine, a barbed wire scrawl the length of the thumb,
the map winter makes on a closed fist. Your gloves
wear me, content again, their small heat
enveloping my hands.

The Boat

BY

ALISTAIR

MACLEOD

~

There are times even now, when I awake at four o'clock in the morning with the terrible fear that I have overslept; when I imagine that my father is waiting for me in the room below the darkened stairs or that the shorebound men are tossing pebbles against my window while blowing their hands and stomping their feet impatiently on the frozen steadfast earth. There are times when I am half out of bed and fumbling for socks and mumbling for words before I realize that I am foolishly alone, that no one waits at the base of the stairs and no boat rides restlessly in the waters by the pier.

At such times only the grey corpses on the overflowing ashtray beside my bed bear witness to the extinction of the latest spark and silently await the crushing out of the most recent of their fellows. And then because I am afraid to be alone with death, I dress rapidly, make a great to-do about clearing my throat, turn on both faucets in the sink and proceed to make loud splashing ineffectual noises. Later I go out and walk the mile to the all-night restaurant.

In the winter it is a very cold walk and there are often tears in my eyes when I arrive. The waitress usually gives a sympathetic little shiver and says, "Boy, it must be really cold out there; you got tears in your eyes."

"Yes," I say, "it sure is; it really is."

And then the three or four of us who are always in such places at such times make uninteresting little protective chit-chat until the dawn reluctantly arrives. Then I swallow the coffee which is always bitter and leave with a great busy rush because by that time I have to worry about being late and whether I have a clean shirt and whether my car will start and about all the other countless things one must worry about when he teaches at a great Midwestern university. And I know then that that day will go by as have all the days of the past ten years, for the call and the voices and the shapes and the boat were not really there in the early morning's darkness and I have all kinds of comforting reality to prove it. They are only shadows and echoes, the animals a child's hands make on the wall by maplight, and the voices from the rain barrel; the cuttings from an old movie made in the black and white of long ago.

I first became conscious of the boat in the same way and at almost the same time that I became aware of the people it supported. My earliest recollection of my father is a view from the floor of gigantic rubber boots and then of being suddenly elevated and having my face pressed against the stubble of his cheek, and of how it tasted of salt and of how he smelled of salt from his red-soled rubber boots to the shaggy whiteness of his hair.

When I was very small, he took me for my first ride in the boat. I rode the half-mile from our house to the wharf on his shoulders and I remember the sound of his rubber boots galumphing along the gravel beach, the tune of the indecent little song he used to sing, and the odour of the salt.

The floor of the boat was permeated with the same odour and in its constancy I was not aware of change. In the harbour we made our little circle and returned. He tied the boat by its painter, fastened the stern to its permanent anchor and lifted me high over his head to the solidity of the wharf. Then he climbed up the little iron ladder that led to the wharf's cap, placed me once more upon his shoulders and galumphed off again.

When we returned to the house everyone made a great fuss over my precocious excursion and asked, "How did you like the boat?" "Were you afraid in the boat?" "Did you cry in the boat?" They repeated "the boat" at the end of all their questions and I knew it must be very important to everyone.

My earliest recollection of my mother is of being along with her in the mornings while my father was away in the boat. She seemed to be always repairing clothes that were "torn in the boat," preparing food "to be eaten in the boat" or looking for "the boat" through our kitchen window which faced upon the sea. When my father returned about noon, she would ask, "Well, how did things go in the boat today?" It was the first question I remember asking: "Well, how did things go in the boat today?" "Well, how did things go in the boat today?"

The boat in our lives was registered at Port Hawkesbury. She was what Nova Scotians called a Cape Island boat and was designed for the small inshore fishermen who sought the lobsters of the spring and the mackerel of summer and later the cod and haddock and hake. She was thirty-two feet long and nine wide, and was powered by an engine from a Chevrolet truck. She had a marine clutch and a high speed reverse gear and was painted light green with the name *Jenny Lynn* stencilled in black letters on her bow and painted on an oblong plate across her stern. Jenny Lynn had been my mother's maiden name and the boat was called after her as another link in the chain of tradition. Most of the boats that berthed at the wharf bore the names of some female member of their owner's household.

I say this now as if I knew it all then. All at once, all about boat dimensions and engines, and as if on the day of my first childish voyage I noticed the difference between a stencilled name and a painted name. But of course it was not that way at all, for I learned it all very slowly and there was not time enough.

I learned first about our house which was one of about fifty which marched around the horseshoe of our harbour and the wharf which was its heart. Some of them were so close to the water that during a storm the sea spray splashed against their windows while others were built farther along the beach as was the case with ours. The houses and their people, like those of the neighbouring towns and villages, were the result of Ireland's discontent and Scotland's Highland Clearances and America's War of Independence. Impulsive emotional Catholic Celts who could not bear to live with England and shrewd determined Protestant Puritans who, in the years after 1776, could not bear to live without.

The most important room in our house was one of those oblong old-fashioned kitchens heated by a wood- and coal-burning stove. Behind the stove was a box of kindlings and beside it a coal scuttle. A heavy wooden table with leaves that expanded or reduced its dimensions stood in the middle of the floor. There were five wooden home-made chairs which had been chipped and hacked by a variety of knives. Against the east wall, opposite the stove, there was a couch which sagged in the middle and had a cushion for a pillow, and above it a shelf which contained matches, tobacco, pencils, odd fish-hooks, bits of twine, and a tin can filled with bills and receipts. The south wall was dominated by a window which faced the sea and on the north there was a five-foot board which bore a variety of clothes hooks and the burdens of each. Beneath the board there was a jumble of odd footwear, mostly of rubber. There was also, on this wall, a barometer, a map of the marine area and a shelf which held a tiny radio. The kitchen was shared by all of us and was a buffer zone between the immaculate order of ten other rooms and the disruptive chaos of the single room that was my father's.

My mother ran her house as her brothers ran their boats. Everything was clean and spotless and in order. She was tall and dark and powerfully energetic. In later years she reminded me of the women of Thomas Hardy, particularly Eustacia Vye, in a physical way. She fed and clothed a family of seven children, making all of the meals and most of the clothes. She grew miraculous gardens and magnificent flowers and raised broods of hens and ducks. She would walk miles on berry-picking expeditions and hoist her skirts to dig for clams when the tide was low. She was fourteen years younger than my father, whom she had married when she was twenty-six and had been a local beauty for a period of ten years. My mother was of the sea as were all of her people, and her horizons were the very literal ones she scanned with her dark and fearless eyes.

Between the kitchen clothes rack and barometer, a door opened into my father's bedroom. It was a room of disorder and disarray. It was as if the wind which so often clamoured about the house succeeded in entering this single room and after whipping it into turmoil stole quietly away to renew its knowing laughter from without.

My father's bed was against the south wall. It always looked rumpled and unmade because he lay on top of it more than he slept within any folds it might have had. Beside it, there was a little brown table. An archaic goose-necked reading light, a battered table radio, a mound of wooden matches, one or two packages of tobacco, a deck of cigarette papers and an overflowing ashtray cluttered its surface. The

Donald Cameron MacKay *Landscape, Herring Cove* c 1950

brown larvae of tobacco shreds and the grey flecks of ash covered both the table and the floor beneath it. The once-varnished surface of the table was disfigured by numerous black scars and gashes inflicted by the neglected burning cigarettes of many years. They had tumbled from the ashtray unnoticed and branded their statements permanently and quietly into the wood until the odour of their burning caused the snuffing out of their lives. At the bed's foot there was a single window which looked upon the sea.

Against the adjacent wall there was a battered bureau and beside it there was a closet which held his single ill-fitting serge suit, the two or three white shirts that strangled him and the square black shoes that pinched. When he took off his more friendly clothes, the heavy woollen sweaters, mitts and socks which my mother knitted for him and the woollen and doeskin shirts, he dumped them unceremoniously on a single chair. If a visitor entered the room while he was lying on the bed, he would be told to throw the clothes on the floor and take their place upon the chair.

Magazines and books covered the bureau and competed with the clothes for domination of the chair. They further overburdened the heroic little table and lay on top of the radio. They filled a baffling and unknowable cave beneath the bed, and in the corner by the bureau they spilled from the walls and grew up from the floor.

The magazines were the most conventional: *Time, Newsweek, Life, Maclean's Family Herald, Reader's Digest.* They were the result of various cut-rate subscriptions or of the gift subscriptions associated with Christmas, "the two whole years for only $3.50."

The books were more varied. There were a few hard-cover magnificents and bygone Book-of-the-Month wonders and some were Christmas or birthday gifts. The majority of them, however, were used paperbacks which came from those second-hand bookstores which advertise in the backs of magazines: "Miscellaneous Used Paperbacks 10¢ Each." At first he sent for them himself, although my mother resented the expense, but in later years they came more and more often from my sisters who had moved to the cities. Especially at first they were very weird and varied. Mickey Spillane and Ernest Haycox vied with Dostoyevsky and Faulkner, and the Penguin Poets edition of Gerard Manley Hopkins arrived in the same box as a little book on sex technique called *Getting the Most Out of Love.* The former had been assiduously annotated by a very fine hand using a very blue-inked fountain pen while the latter had been studied by someone with very large thumbs, the prints of which were still visible in the margins. At the slightest provocation it would open almost automatically to particularly graphic and well-smudged pages.

When he was not in the boat, my father spent most of his time lying on the bed in his socks, the top two buttons of his trousers undone, his discarded shirt on the ever-ready chair and the sleeves of the woollen Stanfield underwear, which he wore both summer and winter, drawn half way up to his elbows. The pillows propped up the whiteness of his head and the goose-necked lamp illuminated the pages in his hands. The cigarettes smoked and smouldered on the ashtray and on the table the radio played constantly, sometimes low and sometimes loud. At midnight and at one, two, three and four, one could sometimes hear the radio, his occasional cough, the rustling thud of a completed book being tossed to the corner heap, or the movement necessitated by his sitting on the edge of the bed to roll the thousandth cigarette. He seemed never to sleep, only to doze, and the light shone constantly from his window to the sea.

My mother despised the room and all it stood for and she had stopped sleeping in it after I was born. She despised disorder in rooms and in houses and in hours and in lives, and she had not read a book since high school. There she had read *Ivanhoe* and considered it a colossal waste of time. Still the room remained, like a solid rock of opposition in the sparkling waters of a clear deep harbour, opening off the kitchen where we really lived our lives, with its door always open and its contents visible to all.

The daughters of the room and of the house were very beautiful. They were tall and willowy like my mother and had her fine facial features set off by the reddish copper-coloured hair that had apparently once been my father's before it turned to white. All of them were very clever in school and helped my mother a great deal about the house. When they were young they sang and were very happy and very nice to me because I was the youngest and the family's only boy.

My father never approved of their playing about the wharf like the other children, and they went there only when my mother sent them on an errand. At such times they almost always overstayed, playing screaming games of tag or hide-and-seek in and about the fishing shanties, the piled traps and tubs of trawl, shouting down to the perch that swam languidly about the wharf's algae-covered piles, or jumping in and out of the boats that tugged gently at their lines. My mother was never uneasy about them at such times, and when her husband criticized her she would say, "Nothing will happen to them there," or "They could be doing worse things in worse places."

By about the ninth or tenth grade my sisters one by one discovered my father's bedroom and then the change would begin. Each would go into the room one morning when he was out. She would go with the ideal hope of imposing order or with the more practical objective of emptying the ashtray, and later she would be found spell-bound by the volume in her hand. My mother's reaction was always abrupt, bordering on the angry. "Take your nose out of that trash and come and do your work," she would say, and once I saw her slap my youngest sister so hard that the print of her hand was scarletly emblazoned upon her daughter's cheek while the broken-spined paperback fluttered uselessly to the floor.

Thereafter my mother would launch a campaign against what she had discovered but could not understand. At times although she was not overly religious she would bring in God to bolster her arguments, saying, "In the next world God will see to those who waste their lives reading useless books when they should be about their work." Or without theological aid, "I would like to know how books help anyone to live a life." If my father were in, she would repeat the remarks louder than necessary, and her voice would carry into his room where he lay upon his bed. His usual reaction was to turn up the volume of the radio, although that action in itself betrayed the success of the initial thrust.

Shortly after my sisters began to read the books, they grew restless and lost interest in darning socks and baking bread, and all of them eventually went to work as summer waitresses in the Sea Food Restaurant. The restaurant was run by a big American concern from Boston and catered to the tourists that flooded the area during July and August. My mother despised the whole operation. She said the restaurant was not run by "our people," and "our people" did not eat there, and that it was run by outsiders for outsiders.

"Who are these people anyway?" she would ask, tossing back her dark hair, "and what do they, though they go about with their cameras for a hundred years, know about the way it is here, and what do they care about me and mine, and why should I care about them?"

She was angry that my sisters should even conceive of working in such a place and more angry when my father made no move to prevent it, and she was worried about herself and about her family and about her life. Sometimes she would say softly to her sisters, "I don't know what's the matter with my girls. It seems none of them are interested in any of the right things." And sometimes there would be bitter savage arguments. One afternoon I was coming in with three mackerel I'd been given at the wharf when I heard her say, "Well I hope you'll be satisfied when they come home knocked up and you'll have had your way."

It was the most savage thing I'd ever heard my mother say. Not just the words but the way she said them, and I stood there in the porch afraid to breathe for what seemed like the years from ten to fifteen, feeling the damp moist mackerel with their silver glassy eyes growing clammy against my leg.

Through the angle in the screen door I saw my father who had been walking into his room wheel around on one of his rubber-booted heels and look at her with his blue eyes flashing like clearest ice beneath the snow that was his hair. His usually ruddy face was drawn and grey, reflecting the exhaustion of a man of sixty-five who had been working in those rubber boots for eleven hours on an August day, and for a fleeting moment I wondered what I would do if he killed my mother while I stood there in the porch with those three foolish mackerel in my hand. Then he turned and went into his room and the radio blared forth the next day's weather forecast and I retreated under the noise and returned again, stamping my feet and slamming the door too loudly to signal my approach. My mother was busy at the stove when I came in, and did not raise her head when I threw the mackerel in a pan. As I looked into my father's room, I said, "Well how did things go in the boat today?" and he replied, "Oh not too badly, all things considered." He was lying on his back and lighting the first cigarette and the radio was talking about the Virginia coast.

All of my sisters made good money on tips. They bought my father an electric razor which he tried to use for a while and they took out even more magazine subscriptions. They bought my mother a great many clothes of the type she was very fond of, the wide-brimmed hats and the brocaded dresses, but she locked them all in trunks and refused to wear any of them.

On one August day my sisters prevailed upon my father to take some of their restaurant customers for an afternoon ride in the boat.

The tourists with their expensive clothes and cameras and sun glasses awkwardly backed down the iron ladder at the wharf's side to where my father waited below, holding the rocking *Jenny Lynn* in snug against the wharf with one hand on the iron ladder and steadying his descending passengers with the other. They tried to look both prim and wind-blown like the girls in the Pepsi-Cola ads and did the best they could, sitting on the thwarts where the newspapers were spread to cover the splattered blood and fish entrails, crowding to one side so that they were in danger of capsizing the boat, taking the inevitable pictures or merely trailing their fingers through the water of their dreams.

All of them liked my father very much and, after he'd brought them back from their circles in the harbour, they invited him to their rented cabins which were located high on a hill overlooking the village to which they were so alien. He proceeded to get very drunk up there with the beautiful view and the strange company and the abundant liquor, and late in the afternoon he began to sing.

I was just approaching the wharf to deliver my mother's summons when he began, and the familiar yet unfamiliar voice that rolled down from the cabins made me feel as I had never felt before in my young life or perhaps as I had always felt without really knowing it, and I was ashamed yet proud, young yet old and saved yet forever lost, and there was nothing I could do to control my legs which trembled nor my eyes which wept for what they could not tell.

The tourists were equipped with tape recorders and my father sang for more than three hours. His voice boomed down the hill and bounced off the surface of the harbour, which was an unearthly blue on that hot August day, and was then reflected to the wharf and the fishing shanties where it was absorbed amidst the men who were baiting their lines for the next day's haul.

He sang all the old sea chanties which had come across from the old world and by which men like him had pulled ropes for generations, and he sang the East Coast sea songs which celebrated the sealing vessels of Northumberland Strait and the long liners of the Grand Banks, and the Anticosti, Sable Island, Grand Manan, Boston Harbor, Nantucket and Block Island. Gradually he shifted to the seemingly unending Gaelic drinking songs with their twenty or more verses and inevitable refrains, and the men in the shanties smiled at the coarseness of some of the verses and at the thought that the singer's immediate audience did not know what they were applauding nor recording to take back to staid old Boston. Later as the sun was setting he switched to the laments and the wild and haunting Gaelic war songs of those spattered Highland ancestors he had never seen, and when his voice ceased, the savage melancholy of three hundred years seemed to hang over the peaceful harbour and the quiet boats and the men leaning in the doorways of their shanties with their cigarettes glowing in the dusk and the women looking to the sea from their open windows with their children in their arms.

When he came home he threw the money he had earned on the kitchen table as he did with all his earnings but my mother refused to touch it and the next day he went with the rest of the men to bait his trawl in the shanties. The tourists came to the door that evening and my mother met them there and told them that her husband was not in although he was lying on the bed only a few feet away with the radio playing and the cigarette upon his lips. She stood in the doorway until they reluctantly went away.

In the winter they sent him a picture which had been taken on the day of the singing. On the back it said, "To Our Ernest Hemingway" and the "Our" was underlined. There was also an accompanying letter telling how much they had enjoyed themselves, how popular the tape was proving and explaining who Ernest Hemingway was. In a way it almost did look like one of those unshaven, taken-in-Cuba pictures of Hemingway. He looked both massive and incongruous in the setting. His bulky fisherman's clothes were too big for the green and white lawn chair in which he sat, and his rubber boots seemed to take up all of the well-clipped grass square. The beach umbrella jarred with his sunburned face and because he had already been singing for some time, his lips which chapped in the winds of spring and burned in the water glare of summer had already cracked in several places, producing tiny flecks of blood at their corners and on the whiteness of his teeth. The bracelets of brass chain which he wore to protect his wrists from chafing seemed abnormally large and his broad leather belt had been slackened and his heavy shirt and underwear were open at the throat revealing an uncultivated wilderness of white chest hair bordering on the semi-controlled stubble of his neck and chin. His blue eyes had looked directly into the camera and his hair was whiter than the two tiny clouds which hung over his left shoulder. The sea was behind him and its immense blue flatness stretched out to touch the arching blueness of the sky. It seemed very far away from him or else he was so much in the foreground that he seemed too big for it.

Each year another of my sisters would read the books and work in the restaurant. Sometimes they would stay out quite late on the hot summer nights and when they came up the stairs my mother would ask them many long and involved questions which they resented and tried to avoid. Before ascending the stairs they would go into my father's room and those of us who waited above could hear them throwing his clothes off the chair before sitting on it or the squeak of the bed as they sat on its edge. Sometimes they would talk to him a long time, the murmur of their voices blending with the music of the radio into a mysterious vapour-like sound which floated softly up the stairs.

I say this again as if it all happened at once and as if all of my sisters were of identical ages and like so many lemmings going into another sea and, again, it was of course not that way at all. Yet go they did, to Boston, to Montreal, to New York with the young men they met during the summers and later married in those far-away cities. The young men were very articulate and handsome and wore fine clothes and drove expensive cars and my sisters, as I said, were very tall and beautiful with their copper-coloured hair and were tired of darning socks and baking bread.

One by one they went. My mother had each of her daughters for fifteen years, then lost them for two and finally forever. None married a fisherman. My mother never accepted any of the young men, for in her eyes they seemed always a combination of the lazy, the effeminate, the dishonest and the unknown. They never seemed to do any physical work and she could not comprehend their luxurious vacations and she did not know whence they came nor who they were. And in the end she did not really care, for they were not of her people and they were not of her sea.

I say this now with a sense of wonder at my own stupidity in thinking I was somehow free and would go on doing well in school and playing and helping in the boat and passing into my early teens while streaks of grey began to appear in my mother's dark hair and my father's rubber boots dragged sometimes on the pebbles of the beach as he trudged home from the wharf. And there were but three of us in the house that had at one time been so loud.

Then during the winter that I was fifteen he seemed to grow old and ill at once. Most of January he lay upon the bed, smoking and reading and listening to the radio while the wind howled about the house and the needle-like snow blistered off the ice-covered harbour and the doors flew out of people's hands if they did not cling to them like death.

In February when the men began overhauling their lobster traps he still did not move, and my mother and I began to knit lobster trap headings in the evenings. The twine was as always very sharp and harsh, and blisters formed upon our thumbs and little paths of blood snaked quietly down between our fingers while the seals that had drifted down from distant Labrador wept and moaned like human children on the ice-floes of the Gulf.

In the daytime my mother's brother who had been my father's partner as long as I could remember also came to work upon the gear. He was a year older than my mother and was tall and dark and the father of twelve children.

By March we were very far behind and although I began to work very hard in the evenings I knew it was not hard enough and that there were but eight weeks left before the opening of the season on May first. And I knew that my mother worried and my uncle was uneasy and that all of our very lives depended on the boat being ready with her gear and two men, by the date of May the first. And I knew then that *David Copperfield* and *The Tempest* and all of those friends I had dearly come to love must really go forever. So I bade them all good-bye.

The night after my first full day at home and after my mother had gone upstairs he called me into his room where I sat upon the chair beside his bed. "You will go back tomorrow,"" he said simply.

I refused then, saying I had made my decision and was satisfied.

"That is no way to make a decision," he said, "and if you are satisfied I am not. It is best that you go back." I was almost angry then and told him as all children do that I wished he would leave me alone and stop telling me what to do.

He looked at me a long time then, lying there on the same bed on which he had fathered me those sixteen years before, fathered me his only son, out of who knew what emotions when he was already fifty-six and his hair had turned to snow. Then he swung his legs over the edge of the squeaking bed and sat facing me and looked into my own dark eyes with his of crystal blue and placed his hand upon my knee. "I am not telling you to do anything," he said softly, "only asking you."

The next morning I returned to school. As I left, my mother followed me to the porch and said, "I never thought a son of mine would choose useless books over the parents that gave him life."

In the weeks that followed he got up rather miraculously and the gear was ready and the *Jenny Lynn* was freshly painted by the last two weeks of April when the ice began to break up and the lonely screaming gulls returned to haunt the silver herring as they flashed within the sea.

On the first day of May the boats raced out as they had always done, laden down almost to the gunwales with their heavy cargoes of traps. They were almost like living things as they plunged through the waters of the spring and manoeuvred between the still floating icebergs of crystal-white and emerald green on their way to the traditional grounds that they sought out every May. And those of us who sat that day in the high school on the hill, discussing the water imagery of Tennyson, watched them as they passed back and forth beneath us until by afternoon the piles of traps which had been stacked upon the wharf were no longer visible but were spread about the bottoms of the sea. And the *Jenny Lynn* went too, all day, with my uncle tall and dark, like a latter-day Tashtego standing at the tiller with his legs wide apart and guiding her deftly between the floating pans of ice and my father in the stern standing in the same way with his hands upon the ropes that lashed the cargo to the deck. And at night my mother asked, "Well, how did things go in the boat today?"

And the spring wore on and the summer came and school ended in the third week of June and the lobster season on July first and I wished that the two things I loved so dearly did not exclude each other in a manner that was so blunt and too clear.

At the conclusion of the lobster season my uncle said he had been offered a berth on a deep sea dragger and had decided to accept. We all knew that he was leaving the *Jenny Lynn* forever and that before the next lobster season he would buy a boat of his own. He was expecting another child and would be supporting fifteen people by the next spring and could not chance my father against the family that he loved.

I joined my father then for the trawling season, and he made no protest and my mother was quite happy. Through the summer we baited the tubs of trawl in the afternoon and set them at sunset and revisited them in the darkness of the early morning. The men would come tramping by our house at four A.M. and we would join them and walk with them to the wharf and be on our way before the sun rose out of the ocean where it seemed to spend the night. If I was not up they would toss pebbles to my window and I would be very embarrassed and tumble downstairs to where my father lay fully clothed atop his bed, reading his book and listening to his radio and smoking his cigarette. When I appeared he would swing off his bed and put on his boots and be instantly ready and then we would take the lunches my mother had prepared the night before and walk off toward the sea. He would make no attempt to wake me himself.

It was in many ways a good summer. There were few storms and we were out almost every day and we lost a minimum of gear and seemed to land a maximum of fish and I tanned dark and brown after the manner of my uncles.

My father did not tan—he never tanned—because of his reddish complexion, and the salt water irritated his skin as it had for sixty years. He burned and reburned over and over again and his lips still cracked so that they bled when he smiled, and his arms, especially the left, still broke out into the oozing salt-water boils as they had ever since as a child I had first watched him soaking and bathing them in a variety of ineffectual solutions. The chafe-preventing bracelets of brass linked chain that all the men wore about their wrists in early spring were his the full season and he shaved but painfully and only once a week.

And I saw then, that summer, many things that I had seen all my life as if for the first time and I thought that perhaps my father had never been intended for a fisherman either physically or mentally. At least not in the manner of my uncles; he had never really loved it. And I remembered that, one evening in his room when we were talking about *David Copperfield*, he had said that he had always wanted to go to the university and I had dismissed it then in the way one dismisses his father's saying he would like to be a tight-rope walker, and we had gone on to talk about the Peggottys and how they loved the sea.

And I thought then to myself that there were many things wrong with all of us and all our lives and I wondered why my father, who was himself an only son, had not married before he was forty and then I wondered why he had. I even thought that perhaps he had had to marry my mother and checked the dates on the flyleaf of the Bible where I learned that my oldest sister had been born a prosaic eleven months after the marriage, and I felt myself then very dirty and debased for my lack of faith and for what I had thought and done.

And then there came into my heart a very great love for my father and I thought it was very much braver to spend a life doing what you really do not want rather than selfishly following forever your own dreams and inclinations. And I knew then that I could never leave him alone to suffer the iron-tipped harpoons which my mother would forever hurl into his soul because he was a failure as a husband and a father who had retained none of his own. And I felt that I had been very small in a little secret place within me and that even the completion of high school was for me a silly shallow selfish dream.

So I told him one night very resolutely and very powerfully that I would remain with him as long as he lived and we would fish the sea together. And he made no protest but only smiled through the cigarette smoke that wreathed his bed and replied, "I hope you will remember what you've said."

The room was now so filled with books as to be almost Dickensian, but he would not allow my mother to move or change them and he continued to read them, sometimes two or three a night. They came with great regularity now, and there were more hard covers, sent by my sisters who had gone so long ago and now seemed so distant and so prosperous, and sent also pictures of small red-haired grandchildren with baseball bats and dolls which he placed upon his bureau and which my mother gazed at wistfully when she thought no one would see. Red-haired grandchildren with baseball bats and dolls who would never know the sea in hatred or in love.

And so we fished through the heat of August and into the cooler days of September when the water was so clear we could almost see the bottom and the white mists rose like delicate ghosts in the early morning dawn. And one day my mother said to me, "You have given added years to his life."

And we fished on into October when it began to roughen and we could no longer risk night sets but took our gear out each morning and returned at the first sign of the squalls; and on into November when we lost three tubs of trawl and the clear blue water turned to a sullen grey and the trochoidal waves rolled rough and high and washed across our bows and decks as we ran within their troughs. We wore heavy sweaters now and the awkward rubber slickers and the heavy woollen mitts which soaked and froze into masses of ice that hung from our wrists like the limbs of gigantic monsters until we thawed them against the exhaust pipe's heat. And almost every day we would leave for home before noon, driven by the blasts of the northwest wind, coating our eyebrows with ice and freezing our eyelids closed as we leaned into a visibility that was hardly there, charting our course from the compass and the sea, running with the waves and between them but never confronting their towering might.

And I stood at the tiller now, on these homeward lunges, stood in the place and in the manner of my uncle, turning to look at my father and to shout over the roar of the engine and the slop of the sea to where he stood in the stern, drenched and dripping with the snow and the salt and the spray and his bushy eyebrows caked in ice. But on November twenty-first, when it seemed we might be making the final run of the season, I turned and he was not there and I knew even in that instant that he would never be again.

On November twenty-first the waves of the grey Atlantic are very very high and the waters are very cold and there are no signposts on the surface of the sea. You cannot tell where you have been five minutes before and in the squalls of snow you cannot see. And it takes longer than you would believe to check a boat that has been running before a gale and turn her ever so carefully in a wide and stupid circle, with timbers creaking and straining, back into the face of storm. And you know that it is useless and that your voice does not carry the length of the boat and that even if you knew the original spot, the relentless waves would carry such a burden perhaps a mile or so by the time you could return. And you know also, the final irony, that your father like your uncles and all the men that form your past, cannot swim a stroke.

The lobster beds off the Cape Breton coast are still very rich and now, from May to July, their offerings are packed in crates of ice, and thundered by the gigantic transport trucks, day and night, through New Glasgow, Amherst, Saint John and Bangor and Portland and into Boston where they are tossed still living into boiling pots of water, their final home.

And though the prices are higher and the competition tighter, the grounds to which the *Jenny Lynn* once went remain untouched and unfished as they have for the last ten years. For if there are not signposts on the sea in storm there are certain ones in calm and the lobster bottoms were distributed in calm before any of us can remember and the grounds my father fished were those his father fished before him and there were others before and before and before. Twice the big boats have come from forty and fifty miles, lured by the promise of the grounds, and strewn the bottom with their traps and twice they have returned to find their buoys cut adrift and their gear lost and destroyed. Twice the Fisheries Officer and the Mounted Police have come and asked many long and involved questions and twice they have received no answers from the men leaning in the doors of their shanties and the women standing at their windows with their children in their arms. Twice they have gone away saying: "There are no legal boundaries in the Marine area"; "No one can own the sea"; "Those grounds don't wait for anyone."

But the men and the women, with my mother dark among them, do not care for what they say, for to them the grounds are sacred and they think they wait for me.

It is not an easy thing to know that your mother lives alone on an inadequate insurance policy and that she is too proud to accept any other aid. And that she looks through her lonely window onto the ice of winter and the hot flat calm of summer and the rolling waves of fall. And that she lies awake in the early morning's darkness when the rubber boots of the men scrunch upon the gravel as they pass beside her house on their way down to the wharf. And she knows that the footsteps never stop, because no man goes from her house, and she alone of all the Lynns has neither son nor son-in-law that walks toward the boat that will take him to sea. And it is not an easy thing to know that your mother looks upon the sea with love and on you with bitterness because the one has been so constant and the other so untrue.

But neither is it easy to know that your father was found on November twenty-eighth, ten miles to the north and wedged between two boulders at the base of the rock-strewn cliffs where he had been hurled and slammed so many many times. His hands were shredded ribbons as were his feet which had lost their boots to the suction of the sea, and his shoulders came apart in our hands when we tried to move him from the rocks. And the fish had eaten his testicles and the gulls had pecked out his eyes and the white-green stubble of his whiskers had continued to grow in death, like the grass on graves, upon the purple, bloated mass that was his face. There was not much left of my father, physically, as he lay there with the brass chains on his wrists and the seaweed in his hair.

My Collection

BY

DINYAR
GODREJ

I remember mother washing mulberries before they went off,
first thing in the morning, to send to the neighbours. Or washing
our clothes in a bucket, whites, then coloureds, then socks
until the soap turned to slime. I remember her
dressing up food from previous meals and saving scraps
for passing strays who might linger to guard our house by night.
Upon our travels we carried little bags of tea, sugar, rice,
to save on bills and returned with whatever was going cheap.
She had me convinced my brother's trousers were fashionable
a year after he'd outgrown them. In summer she'd unravel
  jerseys
and begin to reknit. My father encouraged such thrift
as people who earn in cash do and mother knew there
was no recovering what was laid to waste.
So she gathered, pickled, stewed, cut lights
from old boxes, strewed her flower beds with
fish scales and egg shells and sacrificed
her siesta to shoo parrots from our mango tree.
Once she even made her own shampoo but that
didn't quite work.

I remember her saving insect-ridden flour, spreading
it out in the noonday blaze, until all that crawled
had crawled away. Old sheets turned
into shoe cloths and dusters, and "good, strong
plastic bags" were saviours of her ripening fruit.
As children we didn't think the secret of her prodigious hands
was in a thousand things she'd transformed or stored away.
We took what she had to give and she didn't stint.

I speak with her monthly on the telephone now—
Am I wearing a vest, do I eat sensibly, then
she puts it down, thinking of the money I spend.

You needn't worry, mother.
Everything is saved. Nothing
is thrown away.

Faces

BY

JULIA LOWRY

RUSSELL

ooking into the mirror, Janis saw at a glance everything she hated about her face: She looked just like her mother. The round face that refused to be anything but round; the short pug nose; and the beginnings of a double chin. Ugh! She hated it when anyone said (and someone always did) "Oh, she looks just like you, Margery," and then grinned as if he or she had found the perfect way to flatter both of them. Janis usually just *grinned* back. Who wants to look like her mother!

Taking a jar of cleansing cream out of the medicine cabinet, she lost her reflection for an instant only to have it reappear when she closed the door again. After opening the jar and balancing the lid on the rim of the sink, Janis began slowly and methodically to cover her face and throat with cream.

"You really get to me." Janis cursed her mother's reflection. Her feelings for her mother were a mixture of bitterness and admiration. "Nothing I've ever done in my entire thirty-two years has ever truly satisfied you," she told her. "So why was I surprised tonight when you really liked the blouse I got you for your birthday. 'But pink? You know I never wear pink, Janis. Didn't they have any other colors? A nice blue, perhaps?'" She mimicked her mother's oh-you-shouldn't-have-but-since-you-did-why-didn't-you-do-it-right voice.

"Take that!" Janis playfully slapped a bit of cleansing cream on her mother's rounded cheeks. "And that!" She added a dollop to her pug nose. "Not bad." As she played with the cream and her mother's reflection, she became fascinated by the transformation of their shared visage. The more cream she put on, the more the reflection did not look like her mother and the more mesmerized Janis became. Soon she found herself reaching deep into the jar and scooping out big globs of white cream, putting it thicker and thicker all over her face, creating a new face, a more pleasing mask for her mirror-mother. Soon nothing was left but two big, dark eyes. Janis's brown eyes had never looked like her mother's green, critical orbs. "Green eyes that don't like pink blouses," Janis complained to the two eyes in the mirror.

The dark eyes followed her movements as Janis set the jar in the sink and wiped her hands on the washcloth that she took from the bar beside the basin. She looked up, met the eyes staring out of the cream, and winked. When one of them winked back, she felt a delicious wickedness.

After dropping the cloth into the sink beside the empty jar of cream, she slowly smoothed the prints her fingers had made. Carefully Janis planed the layered cleanser on her mother's face, for it seemed extremely important that not a single line be left. She worked diligently until, even under close scrutiny, she couldn't detect a single trace, just a smooth, white, blank expanse. Still the dark eyes glittered in the mirror, beckoning Janis into deeper mysteries, for, like any mask, the white, molded cream made its revelations and suggested its secrets.

Janis looked at the eyes peering out from the mask. She studied them with some satisfaction. They were definitely hers! She smiled and they smiled in reply, encouraging her. To what? For a long moment they watched each other, Janis and the eyes-that-were-not-her-mother's. Then she knew. She would create the face she had always wanted. "Sculpture in Cold Cream," by Janis Davis.

She lost her mother's reflection once again as she rummaged in the medicine cabinet. There she spotted an old orange stick that had become lost, separated from the manicure set. Perfect. "If cleansing cream can be the clay, then certainly an orange stick can be the awl," she announced out loud, pleased with her ingenuity. After closing the cabinet door, she looked at those eyes and nodded in conspiracy. Janis began another transformation.

Working with increased concentration, Janis moved the top layers of white cream up from the right side of her face, leaving a slightly thinner white covering behind. She mounded the newly moved cream into a thick line that ran a couple of inches under the right eye, starting from the middle of the socket and extending up and around the corner. Here she smoothed it out until it blended back into the rest of the cream that lay along the side and ran only slightly into the edges of her hair. This she accomplished with her finger. But this new mask needed more definition, more refinement. "I *need* an *awl*," she intoned in her best Boris Karloff voice.

With the orange stick she removed some of the roundness from the top of the mound. Carefully she wiped the stick on the washcloth. Then she carved a hollow under the angular mound, a shadow almost. Taking great care, she worked the cream so that no lines or prints appeared on the smooth surface. Surveying her work, Janis liked her artistic endeavors and felt the power of genesis. What is the mystery? Vague shadows flitted through her thoughts, but the eyes smiled their inscrutable insistence and the face with two histories drew her attendance.

Janis acknowledged the duality before her and continued the metamorphosis. First she moved some of the cream from the left lower portion of her mother's mask, creating an angular cheekbone. She made sure the shallow hollow was perfectly formed underneath. Checking to remove any signs of fingerprints pressed into the mask like those left in clay by the ancients, she looked at the reflection. "These cheekbones don't look a thing like my mother's," Janis noted with satisfaction. The eyes in the mirror urged her to complete the creation.

She moved to the nose, feeling an odd thrill of urgency and giddiness. Quickly she rummaged in the cabinet to secure a second jar of cold cream and immediately set to work. Holding the orange stick close to its tip and drawing a line from the inside corner of her right eye down, swerving out a little before she stopped to indicate the nostril opening, Janis watched it emerge. She repeated it on the left side of her face. Good. After putting the stick down, she took both index fingers and smoothed the indentations made by the stick so that her newly defined nose blended gradually into her beautiful high cheekbones. Then, dipping into the sink and into the second jar for more cream, she molded with her fingers a wonderful Roman nose. It seemed to Janis an elegant nose, glorious in its length and flowing lines. Definitely not pug, she thought.

Janis knew that the chin was the next step. Gingerly she molded the cream into a firmly squared jawline and watched the small double chin fade out of prominence.

Janis stopped to admire her work. For a while she reveled in the sharp angles and dramatic shadows, and decided that she was right to hate roundness. The eyes in the mirror smiled their approval, but Janis became irritated with her mother's reflection.

The irritation grew. The more she looked, the more disgruntled she became. Something was wrong with her new face.

It was too smooth. That was it. There were no lines. "Lines reveal character," Janis had heard her mother recite countless times when they discussed their family's ancestry. According to her mother, one could read a person's history in the lines on his face. It was one's ancestors' way of "etching themselves into the future." Janis usually yawned through her mother's family lectures and rarely bothered to credit them with much wisdom. But maybe she was right. For once. (Janis wasn't about to relinquish any more cherished grudges than she had to.) The face etched in cold cream and reflected in the mirror was no longer hers or her mother's. It had no history, no heritage. "This face has no *Mother,* no family, no community."

She removed the washcloth and jars from the bottom of the sink and turned on the taps. One fantasy down the drain, she thought, amused at her own pun. She looked again at the face without a past. "No past means no future," she warned the reflection while she waited for the water to get hot.

Janis stopped her idle movements, arrested by the dark eyes in the mirror that seemed eerie and detached. She peered at the white, grotesque mask. She wasn't smiling anymore, but somehow Janis felt as if the eyes were. She tried consciously to change the expression in her eyes, but nothing seemed to affect them. She shivered. She was uneasy, and a little afraid of those eyes that were watching her, that had teased her and tempted her to mask her mother's face. It dawned on her as they stood facing each other that the game wasn't over.

Bending over the sink, Janis rinsed off the cream and scrubbed her hands, slowly at first and then with more vigor. She wanted to make sure it was all off before she dared look. Groping for a towel, she pressed one hand over her eyes so she wouldn't meet those other reflected orbs. She buried her face into the folds of the towel and waited. One minute. Two. She had to look.

Janis ran her fingers down the length of her nose. It felt right, familiar, not the nose she had created in the mirror. With her right hand she traced the lines of her nose while her left hand followed the cold, elongated beak that extended out and over-shadowed the mouth in the looking glass. The dark eyes in the mirror were filled with horror. Hers? Or theirs?

"No," she murmured quietly. "No...no..." She cupped both hands around her face frantically feeling the round full surfaces that signaled her cheeks, her face. But all the while she saw heinous angles and dark recesses punctuating the mask. In the mirror she saw a severe, sharp jaw that seemed to sever the face from the rest of the body. Eyes glittered and watched in silence.

She panicked. Desperately she ravaged the contours of her face seeking that which the mirror refused to reflect. Where was her face? Where was she in the mirror? Mentally Janis conjured up an old childhood ritual chant: Come out! Come out! Wherever you are! She knew she was somewhere in that mirror, behind that distorted visage. She sought to find herself in the mask, somewhere deep in the creation. She found the face within her hands, but those hideous eyes refused to reflect the truth.

"Come out!" she commanded as her hands began pressing her face with furious motions, trying to gain control of the image. "Come out. Come out. I know where you are," Janis demanded, her hands formed into fists. She pounded her cheeks with heavy blows. "I will come out," she cried with determination to eyes that glared back defiantly and danced with madness to a contrary song. "Please..." she pleaded in a whisper.

Then through the beating of fists and the breaking of glass came the old image. It bounced off the fragmented shards and reverberated from the porcelain. Janis, looking deep within those eyes, saw her mother, aunts, grandmothers, and all the grandmothers before her. She saw that which was within, hidden in the names and faces. Hers and theirs, there in the mirror, reflected again and again and again.

Where My Father Sat

BY

DAVE

MARGOSHES

Where my father sat
the cushions bent
with the weight
of ages and blood
running in his bones
like torrents to drown
the towns
we children built,
crusting foundations
scarred with silt.
The springs groaned
when his texture breathed.

Where my mother sat
the air rose giddily
into the swell of her skirt,
the air was glad.

Grief
and Fear

~

BY

MICHAEL

COREN

An anecdote of the extraordinary within the ordinary. Saturday afternoon in High Park in Toronto. The sun is in one of its relentless moods, inexorable in tossing down furnace-like threads of heat. Lucy holds my hand, tightly. I can feel her pulse, her life. Her diminutive, perfect face is all ice cream and smiles. Four-year-old joy. She plays on the swings, plays on the slide, laughs at the animals in the tiny zoo, skips and jumps and asks me questions that defy philosophy and theology in single, simple bounds. When she runs she frowns and pumps her arms and, like some cartoon character, seems to make hardly any progress at all.

Her red-and-blue ball rolls down an embankment. Only a few feet. I tell my daughter to stand still. I'll get it. Just wait for me, Lucy. There's nobody about. She's safe. I rush down the hill, pick up the ball, rush back again. Only seconds. And Lucy is not there. Lucy is not there. Like submerging your head under water. Every sense—sound and sight and feel and smell and taste—instantly changes its perspective. And a father's sixth sense as well, the one that develops immediately his child is born.

C.S. Lewis said that grief felt like fear. Well, fear also feels like grief. That ancient, emetic feeling that I remember so well just before that fight in the schoolyard, when the biggest kid shouted "I'll see you at 4, outside." Flushed, swallowing, blood racing. The first time I para-chuted, I thought that I would never be as frightened again in my life. I was wrong. Lucy, Lucy! A horrible blending of anger, devotion and fear. I've told her that if she ever loses sight of her mum or dad she must look for a woman or man with children or a police officer and tell them her telephone number. Just seven digits, formal and dull. 920-4500. But not for Lucy. "Nine-two-zero-forty-five hundred" she would chant the number as if it were a panacea-like mantra, and then giggle. She'd say it over and over again so as to tease us, and throw her head back with more laughter. Every time we said "Lucy, that's enough," she would just start again. Monkey.

Now, without thinking, I am reciting it under my breath. Nine-two-zero-forty-five-hundred, nine-two-zero-forty-five-hundred. Reassurance. Like some mindless, numbing repetition of a rosary. And I realize that I am panting, and that I have been sprinting around the park for 15 minutes. I am covered in sweat and I have also cut my leg on a sharp branch. I hadn't noticed; I'm clenching my fists. Say it Lucia, say it, shout it, as loud as the thunder that you love so much when you are safely in your bed, so that I can hear it and come and get you. Nine-two-zero-forty-five-hundred. But there's nothing.

Fear getting stronger now. I'm not making any bloody progress. God, Mike, end this! Time to make a decision. Do I continue this ani-malistic gust or do I retreat to a telephone and call for help, notify the police? I'm usually so definite, so calm in judgment. Now I'm sinking and sinking, Somme-like, in a mud of panic and ambivalence.

And then God proves his existence. Late, as always. Standing behind a tree, so small and so vulnerable, is the puckish child whose cord of life I cut and whose existence is a crucial part of any political or social comprehension. She is crying, because the hiding was a game in the beginning but then she saw how sad and excited I was and thought I would be cross when she stopped the game. I hold her, sur-round her with me, smother her in me. She stops crying and says, "Daddy, you looked so funny." I suppose I did.

An anecdote of the extraordinary within the ordinary. The bond between a father and his child. An everyday event that contains within it aspects of the most significant and precious. Because something is numerically common does not preclude it from being qualitatively spectacular. Let us now define fatherhood, and let us define it with care, passion and understanding.

Learning to Drive

~

BY

MARK

VINZ

My daughter swerves, then struggles with
the wheel, jaw set. She'll get it right
next time, or maybe the next. Turns
are always hardest on these pot-holed streets,
out in the industrial park on Sunday morning.
Don't forget to signal, check the mirror
again, you'll get the hang of it—what
my father told me on some empty
country road, though he never said to
hit the brakes first, just take the wheel
and steer, make a left down at that mailbox.
I scattered gravel for two hundred feet
before we skidded toward the ditch.
Slow down, I say, and watch for ruts.
We drive in Sunday morning circles, lost
in lessons—trying to remember how
you never can predict what's out there,
not even on those routes you know the best.

Now, A Word From Mother

~

BY

MEGAN

ROSENFELD

I was sent to a Capitol Hill press conference at which the major networks announced their plan to label violent television shows. It was a last-minute assignment, and I grabbed the only note pad I could find in the house. Once I sat down in the crowded and tension-filled hearing room in the Dirksen Senate Office Building, I turned to the top page of the note pad and found a drawing by my nine-year-old son—of a beautifully detailed weapon carefully labeled "infinity war mutant bounty gun."

So, guys, I said to myself as I leaned back and looked at the suits lined up at the microphone, tell me about violence on TV.

There has been a lot of moaning in reaction to these labels. One side, the anti-violence lobby, says they're a palliative, a crumb tossed at the mob to stall legislators from actually regulating the airwaves, a license to produce even more violence as long as it's labeled.

The other side, the so-called "creative community" and the civil liberties dogmatists, is crying about artistic freedom and the First Amendment. Labeling, they say, is the first step down the slippery slope of censorship. "I think we are being enormously scapegoated," producer Dick Wolf of *Law and Order* told this newspaper [*The Washington Post*]. If you listen to Congress, it's [as if] television is the root of all social ills in this country."

Oh, please. I find myself curiously unmoved by television producers covering themselves with a First Amendment flag. As far as I'm concerned, they have abrogated their rights to freedom of speech by being so resolutely unconcerned about the impact of what they put on television. That includes the 100 000 acts of violence that the average child will have watched by the end of elementary school. Too often television is not about protecting speech, [but] protecting a business. Pretending otherwise is foolish.

The much-lauded Steven Bochco's series on ABC, *NYPD Blue*, has been widely predicted to "push the envelope" of profanity, nudity and artistic violence. What happens when he gets bored with the show, as he did with *L.A. Law,* and leaves it in the hands of lesser mortals? We'll be left with a flabby envelope and no art.

An Unfortunate Tradition

We have an unfortunate tradition in this country of putting the "rights" of adults before the needs of children. You can see that demonstrated every day in courthouses, classrooms and social work bureaucracies everywhere. Television is no exception.

Every socially useful change that has been made in recent years—like the elimination of cigarette and liquor advertising or identifying the early evening hours for "family" viewing—has been forced down the networks'

throats. Somehow, I am not convinced that they want to use the First Amendment in my family's interest.

"The whole issue is very circular," Marcy Kelly, president of a Los Angeles-based watchdog and education group called Mediascope, told *CQ Researcher.* "The writer will say, 'That's what the producer wants me to write.' The producer will say, 'That's what the network wants.' The network will say, 'That's what the advertisers want.' The advertisers will say, 'That's what the public wants.' The question is, how can we break the circle?"

Well, I have a few ideas.

Misreading the Viewer

First, the notion that blood is what the public wants is not borne out by the facts. The 20 top-rated network shows include only two—*Rescue 911* and *Unsolved Mysteries*—that could conceivably be classified as violent. Most of the top-rated shows are sitcoms or newsmagazines— in other words, just programs that consist of tasteless jokes, sexual and racial stereotypes, and the occasional thoughtful, well-researched exposé. But not violence.

A Times-Mirror survey of more than 1500 adults released in March 1993 found that 80 percent thought that television violence—whether on networks or cable—was harmful to society, compared with 64 percent who thought so in 1983. The number who thought it was "very harmful" (as opposed to just plain harmful) increased from 26 percent to 47 percent. The most frequent viewers of violent programs were those under the age of 30....

Parent's Prerogative

Whenever I interview people in the film or television business I always ask them if they restrict or monitor their child's television watching. They always say they do. Steven Spielberg, for example, said he tried to control what his children watch, although, he admitted, "not enough." And when asked about the question of violence on television, he said, "There's nothing we can do about it except continue to be critical. To protect our children we must be hypocrites."

I guess that's me.

I am not a person who wants to eradicate all trash on television. Nor do I think Looney Tunes should be listed among the most violent shows, as the National Coalition on Television Violence does. I have friends, and even relatives, who write for and act on television, and other friends who would love to do so. I am happy to assert my responsibility as a parent.... I know that little boys will draw pictures of guns, and always have drawn pictures of guns, regardless of how much violence they see on television. And I do not think a television show causes someone to go out and commit murder; it just gives them an excuse after the fact.

But I admit it. If there were such a thing as the Taste Police, I would want to be the chief. We all think we know what's best, especially when it comes to our children. And tell me, have the presidents of NBC, CBS, ABC and the schlock specialists at Fox Broadcasting done such a great job of looking out for our interests?

Mastering the Television

I have always believed that television is a tool you should teach your children to master, not avoid, because it is unavoidable. Throwing it out is a solution, but that doesn't teach children to make choices. In my household, we don't allow television on school nights, for example, and we try to limit it at other times. But the pull of the black box in the corner is always there, and rare is the day that my kids don't ask, "Can I watch TV?" And then, "Why not?"

I could use some help. I feel as if I spend too much time as a parent censoring and saying no, and fighting off the unceasing lures thrown out by the [CBS President] Howard Stringers of the world.

A technological aid—a device that would allow me to block out programs I expect to be inappropriate—fills me with a combination of dread and curiosity. I will never buy those 500 channels we are being promised. But spending some of my rare non-working hours programming a week's worth of television is not my idea of quality time.

A call to Circuit City elicited the information that I can now buy a television with a "parent control capability." I could program when it would turn on and off and block out some channels. This would cost me only between $600 and $700. Perhaps Steven Bochco would like to buy me one.

My stepson told me someone has invented a bicycle-powered television, which would certainly cut down on viewing in our household. I'm not sure it really deals with the violence question, however.

So, for now, I'll take the labels. I would prefer that the creators of television would produce less violence, but perhaps this initiative will help make that happen. It is, at least, an acknowledgement that what they put on the air really does affect human life, and for that I am grateful. Previously, network executives sold advertising on the premise that commercials would prod us lowly viewers to buy products, but refused to admit that the programs could possibly influence anyone to do anything.

I hope the industry, including the cable operators and independent producers who refused to sign on to the two-year labeling experiment, will take seriously the companion plan to sensitize producers, writers and program developers to violence. The objective is to do for violence what they did for seat belts, drug use and cigarette and alcohol consumption—glamorize the good stuff and downplay the bad.

There is a real possibility that nothing will change. In that case I will have no choice: I will have to take the television out and throw a bomb at it. I sure hope nobody else will be hit by the flying debris.

A Sweet Sad Turning of the Tide

BY

**MAUDE
MEEHAN**

My two young sons
move with assurance
through the maze
of ropes and sails,
steer out of turbulence
to calmer seas,
drop anchor.

They climb the mast;
my body tenses
with past apprehension.
Suddenly one dives,
I plummet with him,
breathe again
when he emerges.

The small boat pitches
as he hoists aboard.
I glance up swiftly
to the swaying crossbeam
where his brother
perches, confident.

As heavy mist rolls in
they guide the light craft
back to harbor, and like my hands
old maps lie folded in my lap.

Reaching the laddered dock
they stretch strong arms
to steady me. When did
this turnabout occur?
I have become a passenger
on their journey.

Standard Answers

~

BY

PEG

KEHRET

Cast: One **Parent,** *four* **Students.**
No setting necessary.
At Rise: The four **Students** *are seated on the floor.* **Parent,** *standing,*
addresses the audience.

. . . .

Parent: I am Parent. I am Author. All other parents in the universe use my book as their guide to raising children. They use it now; they used it a hundred years ago; they will use it in the future. It is timeless and ageless.

It is the best-kept secret ever. All parents know about it; kids never know, until they grow up and become parents themselves, and by then, they are ready to use my book and so they willingly perpetuate the secret, keeping it going throughout infinity.

What book? you ask. All right. I will share the secret, even though some of you in the audience are not yet parents yourselves. I am the author of *The Parents' Book of Standard Answers.* You may have wondered why it is that parents in New York City and parents in Taos, New Mexico, will answer certain questions exactly the same way. It is because they all use my book.

To demonstrate the remarkable universality of my book, I have asked three students to assist me. (Three **Students** stand.) No matter what they say to me, my replies will come from *The Parents' Book of Standard Answers.*

Student One: I studied for that test. I don't know why I failed it.

Parent: You didn't study hard enough.

Student Two: I'm full.

Parent: Eat your vegetables, or you don't get dessert.

Student Three: Why can't I go? Give me one good reason.

Parent: Because I said so.

Student One: It isn't fair.

Parent: Life isn't always fair.

Student Two: All the other kids are doing it.

Parent: I don't care what the other kids do. You are not the other kids.

Student Three: Can I go now?

Parent: *May* I go now?

Student One: Can I take the car to school today?

Parent: What's wrong with the bus?

Student Two: Can I take the car to school today?

Parent: When I was your age, I used to walk two miles to school every day.

Student Three: Don't you trust me?

Parent: I remember what I was like at your age.

Student One: Why do I have to clean my room? It just gets dirty again.

Parent: If you don't clean your room, you'll have spider nests in it.

Student Two: I think I'll quit school and get a job.

Parent: You'll be sorry.

Student Three: I want to make some money so I can buy a car.

Parent: Something wrong with your legs?

Student One: Someday I'm leaving this hick town. I'll travel all over the world.

Parent: There's no place like home.

Student Two: You're mean.

Parent: It's for your own good.

Student Three: Why can't I go? Give me one good reason.

Parent: Someday, when you're my age, you'll understand.

Student One: I didn't ask to be born.

Parent: Well, it's too late now.

Student Two: There's nothing to eat around here.

Parent: Have an apple.

Student Three: There's nothing to do around here.

Parent: Read a book.

Student One: I'm bored.

Parent: Mow the lawn.

*(Fourth **Student** leaps to his or her feet and shouts:)*

Student Four: I am Child. I am Author. I also wrote a book that has been used for decades. Every child intuitively knows my book; it is part of the wisdom of the ages, which is passed along on the wind. My book is *The Kids' Book of Standard Answers.*

Parent: What are you doing?

Student Four: Nothing.

Parent: Don't interrupt.

Student Four: I'm not interrupting. I'm joining the conversation.

Parent: We'll discuss this later.

Student Four: I want to discuss it now.

Parent: This isn't a good time for it.

Student Four: Why not?

Parent: Drink your milk.

Student Four: I'm full.

Parent: Who ate all the cookies?

Student Four: Not me.

Parent: Who tracked mud on the carpet?

Student Four: I don't know.

Parent: Did you hear what I said?

Student Four: Huh?

Parent: You kids are driving me crazy with your fighting.

Student Four: He hit me first.

Parent: You've been on the phone for forty-five minutes. Someone may be trying to call.

Student Four: Who?

Parent: I don't know who, but it's time for you to hang up.

Student Four: I'm not done talking.

Parent: Brussels sprouts are good for you.

Student Four: Yuck.

Parent: This concludes our demonstration.

Student One: Wait a minute. I'm not finished.

Student Two: Me, either. I want to say more.

Student Three: Why do we have to stop now?

Parent: Because I said so. *(Four **Students** look at each other, shrug helplessly, and exit, followed by **Parent**.)*

April 19th, 1985

~

BY

**DEIRDRE
LEVINSON**

s we enter the park this forty-second anniversary of the Warsaw
Ghetto uprising, his eyes on a passing cyclist's new-fangled
mount, he starts in on the evergreen topic of his next birthday,
he bets we can't guess what he wants for it—talking fast against our
full-throated protest that he's got ten months to go yet—he'll give us a
clue, it begins with B. There was a time I would not have believed it
possible for any child of mine to show himself on so grave an occasion
so lightminded. But that was before I had children, before my old compulsion to raise them as living monuments to our pulverized people
was foundered by the children themselves. So all I say, as we dispose
ourselves for a while beside the flower-beds on the promenade, is that
if it's birthdays we're talking about it had better be hers.

She will be fifteen imminently. Not a fortnight to go, but she still
hasn't decided what she wants for this birthday, she says she just can't
get excited about this one. For my fifteenth birthday I got new school
underwear—navy-blue flannelette, with elastic at the leg—from my
mother, and a Hebrew prayer-book and five bob from my father. "And
a rich haul I thought it too," I add, with improving intent.

"Mercifully," the chit observes, "times have changed since those
Stonehenge days. Don't tell us what you bought with the five bob, we

know what you bought. Grandma told us; she's always telling us. You bought a big fat poetry-book, then you learnt all the poems in it by heart. Wonder-child!" she pinches my cheek. "My wonder-child mommy, I don't want a big fat poetry-book to learn by heart for my birthday."

"Me neither," he echoes feelingly.

"Poems you learn by heart at your age," I soldier on, "you remember for life. You'll always have them with you for company, speaking to you, sustaining you, they'll be your friends in the unfriendliest times."

He begs to differ, urging the livelier merits of a dog for company any old time. But she takes a high tone. "I'm here to tell you that I don't plan on having *poems* for friends. Whatever I do, wherever I go, there'll always be people for me to make friends with. Just regular human people, that's all, that's good enough for this human person."

Arts disqualified, letters dismissed. Just regular human people, that's all, that's good enough! How to expose this meretricious pop-culture humanism, this ass in a lion's skin, for the imposture it is?

"Now don't go taking it *personally*," she gives me a consolatory little kiss, as the two of them move off together towards the playground. "See you later, Mater. I must just go see if my babies are playing there today."

Her babies, twins, live in our building. Their mother is always telling me what a treasure, how indispensable a part of their household, our daughter is, how unfailingly reliable, understanding, sentient. "A wise one, you know what I mean, you can talk to her about anything. I call her my best friend."

You can talk to her about anything, anything uncomplicated by booklearning. Her intellectual life stops at school. She is an aficionado of the popular culture. Her ear is attuned to pop music exclusively; she has papered her room with pictures of rock stars; she has to be dragged by the hair, her gorgeous honey-brown haystack of hair, to an art museum, unless school requires it; she reads nothing, outside of her school texts, but pulp. What sort of sustenance is that to uphold her when it comes to the crunch? As for him, bless his mettlesome spirit and sociable ways, but abandon all hope of an elevated mind from that quarter. Ask him what he wants to be when he grows up, he says rich.

Here they come now, both pushing a double stroller containing the twins, whose mother hasn't waited to be asked twice. We walk briskly along the pedestrian path to the memorial stone, by the time we reach which, the small number assembled there—some old Europeans, survivors, and a handful of young people, their heirs and repositories of their story—are already singing. There are some short speeches to follow. One of the speakers urges us to remember also the gentile dissenters who died in the camps, as we salute the memory of our people who fought to the death in the Warsaw and Vilna ghettos. Then we sing a final anthem in Yiddish, and a couple of old women throw flowers over the iron railings onto the memorial stone. They give a flower each to the children. She picks up one twin, he the other, and they all drop their flowers onto the stone. Standing there with the children, I think of the nameless father briefly mentioned in an affidavit on one of the Ostland massacres. He was standing, the report relates, his young son beside him, on the edge of the mass grave they were shortly to lie in, stroking the child's head as he pointed to the sky, and seemed to explain something to him.

On our way home, retailing the scene to the children. "If you were that father, what would you be saying to your little boy?" I ask my little boy. But it isn't the father he's thinking of. "I'd be peeing in my pants," he says.

"Supposing it were you with us," before I can put my question to her, she returns it adroitly to where it belongs, "what would you say to us?"

Answer that. What would I say to them, pointing to the sky? Would there still be a sky? What to say that could have any meaning for them? I stand at the edge, she on one side, he on the other, the guns at our backs, the dead at our feet. "I'd tell you what a privilege it's been to bring you up. I'd thank you with all my heart for being the children you are," I say in a rush.

"Do you mean it, Ma?" they ask together, astonished, but no more astonished than I am. Then he says, "Can we please change the subject now, Ma? This one gives me the spookies." But the chit, leaving him to wheel the babies, holds my hand all the way home.

from

Hamlet, Act I, Scene III

BY WILLIAM SHAKESPEARE

Polonius to his son, Laertes.

There; my blessing with thee!
And these few precepts in thy memory
Look thou character. Give thy thoughts no tongue,
Nor any unproportioned thought his act.
Be thou familiar, but by no means vulgar.
Those friends thou hast, and their adoption tried,
Grapple them to thy soul with hoops of steel.
But do not dull thy palm with entertainment
Of each new-hatched, unfledged comrade. Beware
Of entrance to a quarrel. But, being in,
Bear 't that the opposed may beware of thee.
Give every man thine ear, but few thy voice.
Take each man's censure, but reserve thy judgment.
Costly thy habit as thy purse can buy,
But not expressed in fancy; rich, not gaudy;
For the apparel oft proclaims the man,
And they in France of the best rank and station
Are of a most select and generous chief in that.
Neither a borrower nor a lender be,
For loan oft loses both itself and friend,
And borrowing dulls the edge of husbandry.
This above all: to thine own self be true,
And it must follow, as the night the day,
Thou canst not then be false to any man.
Farewell; my blessing season this in thee!

January
Chance

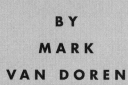

BY

MARK

VAN DOREN

All afternoon before them, father and boy,
In a plush well, with winter sounding past:
In the warm cubicle between two high
Seat backs that slumber, voyaging the vast.

All afternoon to open the deep things
That long have waited, suitably unsaid.
Now one of them is older, and the other's
Art at last has audience; has head,

Has heart to take it in. It is the time.
Begin, says winter, howling through the pane.
Begin, the seat back bumps: what safer hour
Than this, within a somnolent loud train,

A prison where the corridors slide on
As the walls creak, remembering downgrade?
Begin. But with a smile the father slumps
And sleeps. And so the man is never made.

Pockets

BY

KAREN

SWENSON

The point of clothes was line,
a shallow fall of cotton over childish hips
or a coat ruled sharply, shoulder to hem,

but that line was marred by hands
and all the most amazing things
that traveled in them to one's pockets
goitering the shape of grace with gifts—

a puffball only slightly burst
five links of watch chain passed secretly in class
a scrap of fur almost as soft as one's own skin.

Offended at my pouching of her Singer stitch
my mother sewed my pockets up
with an overcast tight as her mouth
forbidding all but the line.

I've lived for years in her seams—
falls of fabric smooth as slide rules
my hands exposed and folded from all gifts.

And it is only recently, with raw fingers
which still recall the warmth and texture of presents,
that I've plucked out stitches sharp as urchin spines
to find both hands and pockets empty.

Our Eldest Son Calls Home

BY

DON

POLSON

To-night you break a week's still absence
to assure us you are safe and well.
Still the father I was years ago,
I am unable to neglect the shadow of
the child I fear, at times, may haunt you.
But though I strain for nuances of loneliness
regret, your voice demands our confidence.
So, for the moment, I am relieved—
unlike that monstrous day I recollect:

Crimson in infantile rage,
our eyes briefly negligent, you stiffened,
turned and fell from the open crib.
Then, like a lifeless doll, mother crying
"My God, he's dead!," we swept you up
into our arms and ran, abandoning the house.
Half-naked and insane with grief and guilt,
we screamed our way through plodding traffic.
Only the next day when they pronounced you
wonderously whole could we deride our frenzy.

Now, beyond the wooded lakes and
distant prairies, you're telling us
of climbing high onto those cool alluring slopes
where you've observed small nestling eagles.
Prayerfully I listen, simply ask you
write us soon and just be careful
—sense the aging of my arms.

I'm Not Slave Material

BY

**ELLEN
GILCHRIST**

"Your children are not your children. They are the sons and daughters of life's longing for itself," wrote Kahlil Gibran. How he knew this is marvelous to me, since I have striven so long and hard to learn it. The rules I discovered while raising my own three sons are simple. You have to let them go. You have to let them ride off on their bikes when they are seven and you have to stop giving them money and advice when they are 21. Of course, as long as you are giving them money you will think you have the right to give them advice. They will resent the advice. If you raised them, it won't be a thing they haven't thought of for themselves.

Nothing that you say now will help. They have to find their manhood. They have to forge their sword. Alone in the forest, like Siegfried, with only the broken pieces of his father's sword for material, they must forge the weapon they will use to kill the dragons of the world. It is very difficult to believe that they are capable of doing this. You must try to see the face of a grown man before you, not the face of the child you taught to tie his shoelaces.

I make it a point to live far away from both my parents and my children. If I lived near my parents I would be sucked into the morass of family problems which is the bread and water on which my parents thrive. They have partaken of this food for so long they think it's nectar. On the other hand, if I lived near my grandchildren I would become their slave. And I'm not slave material.

I believe that the greatest gift I can give my children is to have a life of my own that is not dependent on them. When they come to visit, they come of their own free will, not because they think I need them.

I try not to ask my children too many questions. Every question that a parent asks a child is a leading question. I try not to compare them to other people. And I am trying to learn not to be proud of them. The thing I am proud of today may be the thing they need to stop doing tomorrow in order to grow and change.

I am trying to learn to love them unconditionally, and this is where the money comes in. Giving money to a grown child is giving him power without responsibility. It will keep you from ever having a peer relationship with your child. A relationship where nobody gives advice or hovers or nags. In a peer relationship two people of equal worth and intelligence discuss the world *outside themselves*. They meet as equals and part without sorrow, trusting each other to return with good news. If I'm not paying for it, if it isn't going to cost me money, I can listen to their wildest plans with enthusiasm and respect.

Most of all I want to learn to trust in the future, which is always full of surprises. I am fiercely independent. So is my father. He is 85 years old, and last month, for the first time in our lives, I was able to do something for him that he couldn't do for himself. I went to visit my parents and found him trying to read a book with a plastic sheet of magnifying material he had ordered from a magazine. He is a constant reader. If he isn't working or plotting to control somebody's life, he is reading. In the last few years much of his reading has been about groups who are plotting to control the world. What he reads is not my business. His failing vision is another matter. He had just received word from a third doctor that it was impossible to operate and let more light into his eyes. So he was struggling away with the magnifying sheet and not complaining.

Luckily I had known a lawyer with a similar problem and had seen the marvelous magnifying instruments his wife had found for him.

I left the house on some flimsy excuse and spent the afternoon combing Jackson, Mississippi, for magnifying devices. Finally, in a mall, in a German instruments store, I found it. A raised, lighted rectangle that sat upon the page and made up for the light lost to the retina. I took it home and gave it to him. It was a great moment for me. Later last month, in Paris, I found an even better device that fits into the palm of the hand and can be moved more easily across the page. I had it wrapped and shipped to him with a bottle of Guerlain's L'Heure Bleue for my mother.

There was a message on my answering machine from him when I came home last night. In his kind, cultured, vastly humorous, sweet Southern voice it said, "Sister, I was just calling to thank you for that nice, sophisticated reader you got, and I liked the perfume I was putting on myself too. Ten, Four."

A long time ago we had metaphors for all that I have said. Push them out of the nest. Cut the apron strings. These are vast metaphors. Even more important now that we live such long lives. Plenty of time to create real relationships with the ones we kicked out, or the ones we wish had kicked us out a whole lot sooner.

Dear Mom: This Is the Letter I Would Have Written

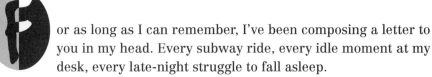

BY

MONICA

STEEL

or as long as I can remember, I've been composing a letter to you in my head. Every subway ride, every idle moment at my desk, every late-night struggle to fall asleep.

Over time, the letters have changed. First they were anonymous notes from caring friends of your children. Next they were brave and desperate heart-songs from me, that I would write just before I went on an extended trip, so you couldn't find me and confirm the demise of our relationship I feared I would have precipitated. But now it's the real thing. Now it's scared and lonely words scrawled, still in utter disbelief, still in absolute confusion, coming from a place no one ever wants to visit in themselves, so wrapped in anger and hatred and betrayal and the most piercingly true love you can imagine.

Now the letter comes because it has to, because I've been advised it might already be too late, because you're an alcoholic and our whole family has so closely followed the classic model so far that my heart stops to think of what the next stops on the map are.

I've spent a lot of the last few months camping out in the "addiction" sections of the bookstores. Vast minutes will pass and I'll realize I haven't swallowed as I read about our lives on the pages. I found a book the other day that totally rejected the concept that alcoholism is a disease. That notion, it argued, frees the drinker of any blame. To do that, I agree, is absolutely unacceptable.

You're not stupid. You're the cleverest person I know. You're perceptive to a fault about the rest of us—how could you not also be so with yourself? That's the part I don't get. We've caught you drinking or drunk a thousand times over the years and, as much as you guiltily crammed those moments to the back of your rum-sodden brain, you knew we knew.

And you kept right on drinking.

When things got bad, or when you wanted to escape—and you almost always seemed to want to escape from something—or now, as the experts tell us, because you have a biological need to do so.

Here's the part I don't get. They say you'll keep on drinking until you reach a crisis point, like running over a child or burning down the house. They say the more we confront and attack you, the more you'll hate yourself and the more you'll drink to deal with that. They say there's nothing at all we could say to you to make you stop, and Dad says he's been told that if you don't, you'll live just five more years.

God, Mom. You'll just be 56. When you were 12, skipping on the sidewalk and watching for Santa from your best friend's window, did you think you'd die a drunk before you even hit old age?

So here's the part I don't get. In your sober moments, surely you must realize what you're doing. We've certainly told you enough times how much it hurts us. *Don't you care?* You're not one of those textbook cases. You're my mom.

How come you hate yourself so much?

People are scared you're going to commit suicide. But you're my mom. You're the one who lifted me off the playground floor when I knocked my teeth on the jungle gym. You're the one who brought home a guinea pig from the zoo for me; the one who wrapped me in her arms when my pet died. You're the one who let us fill the grocery cart with ridiculous things like king-size boxes of maxi pads and cans of Spam.

You're my mom. *It's still you in there, somewhere, I know it is.*

You can't be a textbook case because that would mean you've got years to go before you find the courage to stop drinking, years more of hate and ugliness and loneliness and weakness, and, anyway, you'll probably die before then.

That's the part I don't get. If you really love us, if you really, really do, why won't you stop? And if confronting you just makes you drink more, well, give me some help: What can we do to make you stop? Why won't you give us our wonderful, funny, pretty mother back?

This is the letter I would have written. And now this is the letter I have to write because every time I call and find you lost in that ugly haze that has become your disgusting world, I lose another wonderful, mommy-scented memory. I want my children to have a grandma. I want them to know you as you were, before you succumbed, before you decided your life was so awful you couldn't face it sober.

Please teach me a lesson, Mom. Please teach me about will power and about strength of spirit. Please teach me that nobody's life is perfect and that everybody has to rise above the imperfections and find their own peace. Show me that that peace isn't at the bottom of a bottle. I am your daughter and I need to know that you have these strengths so I can identify them in myself.

And if that doesn't work as an incentive, then maybe consider doing it for you. There's absolutely nothing, absolutely zero positive that comes out of what you're doing right now. You've lost your self-esteem and your dignity and your personality. You're going to die from this and leave two devastated children and a very lonely husband.

So why don't you stop? That's the part I don't get.

I love you very much. But I honestly don't know if you have any love left for me.

Note: *Monica Steel is a pseudonym.*

Ben:
A Monologue

BY

ROGER

KARSHNER

*My mother and father were divorced two years ago and because of
my father's drinking problem I live with my mom. She's a salesperson
for this big clothing manufacturer in New York and her job requires her
to travel a lot. She covers three states—Ohio, Michigan, and Illinois.*

*At first I thought it was neat that she was gone a lot. I mean, I get
to be here alone, I can do what I want, I can stay up late, hang out,
have friends in, whatever. I have total freedom. For a young guy like
me it has to be like this perfect situation, right? Like this dream life.
And all of my friends envy me. They tell me all the time how I've got it
made, how lucky I am, you know. They say that they'd love for their
parents to take off and leave them home alone. Then they could party
all they wanted and not have to do a lot of stupid regimented crap.
They all wish they had the freedom I do.*

*But more and more, as time goes on, I'm starting to think maybe
they're the ones who have it made, not me. When I go to their homes it
feels different. They feel like...like...they feel like* homes. *Here, to be
honest, it doesn't feel like anything. It's just this empty, vacant, nothing. It's just a...just a big nothing place. I mean, yeah, it's a nice*

house, furnished nice and all that, sure. But it doesn't have any warmth, any feelings. At least, when my parents were together, even though they got into major hassles, at least then it felt like a home.

You know, after a while, all the partying and hanging out, the freedom isn't so cool anymore. I mean, how long can you goof off and party? Not that I'm against freedom. No way, not at all. Freedom's important. Freedom is part of us, one of the greatest things in life. But when there's freedom without sharing, without caring, what the hell's the good of it, anyway?

I guess what I'm saying here is that a lot of the time I'm lonely. So lonely that I actually ache inside. Yeah, I really do, sometimes I actually hurt from loneliness. Sometimes when I come home to this empty place, or after a party, after everyone has gone, I get super down, super depressed. And a guy my age shouldn't feel lonely, and empty, and depressed, you know. Hey, it's not right.

I know that my lifestyle looks good to my friends, but they don't realize; they don't realize that they're the ones who have it made. I may have this so-called freedom, but they have a family. They have people around who care.

Strangers

BY

MICHAEL

NGO

n *The Joy Luck Club,* Jing-Mei says about her mother: "We never really understood one another." This is exactly the way I feel about my mother. Every day I lose more and more of my Chinese thinking. It's getting harder and harder to live with my mother.

We cannot communicate with one another. We hardly talk because I have trouble expressing myself. I fully understand what she is saying to me, but I can't answer back. One time we were watching the news together, and she asked me what it was about. So I started explaining in Chinese, and then when she couldn't understand me, I switched to Burmese and still she couldn't understand; so I switched to English hoping she would understand, but I had no luck. Eventually I got so frustrated that I ignored her. Talking with her reminds me of the anger and frustration I felt the first year in Canada when I tried to talk English. Now I hardly ever start a conversation with her.

Sometimes I regret that my English is so fluent. Maybe I should have taken a Saturday morning class in Chinese so I wouldn't have this problem. My friends and I have been here almost the same time, and they don't have trouble speaking with their parents. Their English may be a bit weak, but they are in a better position than I am at home. The way I look at it is that you can't be good in English and your native language as well. It seems as if I had to give up my native language in order to be fluent in English.

As communication disintegrates between me and my mom, I feel more and more separated from her. She has no idea what I want and who I want to be. Sometimes I feel that we're just strangers living together and not a family. My mother, on the other hand, does not see it this way. She thinks that I'm just disobedient. She does not see that I have a problem communicating with her. She thinks that I just don't want to talk to her. Many times she complains that I am disrespectful. I never do what she says and I'm not one of those typically good sons she sees in Chinese movies. She wants someone who is dependent, respectful, and caring—someone like one of my friends who has just arrived in Canada. At first I wanted to be like them—caring and respectful. I really tried, but I realized that I can't because I have grown up in Canada. I am influenced by Canadian culture to be independent and strong. Of course this is the opposite of what my mother wants.

Actually my mother is an old-fashioned woman. She is very superstitious. She still thinks that my two sisters, who died in a car accident, were cursed. So every day, she makes my brother and me carry a bag containing God knows what that will protect us from evil. I refused at first to carry it because not only does the bag bulge out from my pants, but I don't believe in this superstition. However, she scolded me and said it was for my own good and so I had no choice. Often on special occasions we have religious ceremonies which require a lot of work and time. Often I don't bother to participate and again she scolds me, complaining that I'm insulting Chinese culture by not believing in Buddhism.

My mom and I have totally different thoughts. Sometimes I wonder why. Is it because of my personality or because of the Canadian environment I grew up in? My father died in Burma when I was eight years old. Two years later we came to Canada. So I practically grew up here, accepted the Canadian culture, and rejected my Chinese self because I wanted to fit in. Maybe this is the cause of our different values and beliefs. I was too busy trying to gain the benefits of being a Canadian so that I didn't know I was losing myself as a Chinese.

So as every day goes by, I feel ashamed—ashamed that I'm no longer Chinese. I remember one time when I was working, an old

Chinese man asked me a question in weak English. When I couldn't understand him he spoke in Chinese. Knowing I couldn't understand, I asked one of my friends to help him but he was busy so there I was with the old man who was getting more impatient by the minute. Then he finally said, "You are Chinese and you don't speak Chinese."

This really insulted me and I said, "You are in Canada and you don't speak English!" This offended him and he left. Though I was victorious, I felt defeated because what he said really hurt me. I felt ashamed and afraid.

Who am I? I'm Chinese. If so, why don't I think like one? I'm Canadian. But I don't look like one. Who am I? I feel lost—lost between two cultures. Increasingly I feel angry at my mother for forcing me to be what I'm not. You have to let go, Mom, because you lost me the minute I landed on Canadian ground. The harder you try to force me to be what I'm not, the more I will reject my Chinese self. Using force will only make you unhappy and tear me apart. Just let go. You can't have an obedient Chinese son because I have become deeply rooted in Canadian culture. Please, just let go. I don't belong to you anymore.

A
Villager's
Response

~

BY

STEVE

OWAD

esterday, my son came home to Thailand to visit. I am certain it
will be a fine visit for I have not seen him and he has not seen
me or this village in four years. He now lives in a city in Canada,
with modern and fashionable offerings and I know he is happy there.
He only returns to the dried plains of Aranyaprathet to appease my
will. He knows that I often think of him and of how he has chosen to
live so far away, and he knows that such thoughts strike me with a
great sadness. I am an old man now, and since my wife has long since
died and I have no other children, I worry about what will come in the
future.

What will come of my fields, these silent providers and sustainers
for so many generations of family, when they are loosed from my
grasp by death? They will not be inherited by my son, for he does not
want them. They, along with generations of mourning ghosts that had
each, in their time, been their nurturers, will be trampled by the
unsoothing feet of another family. I shall be forced to leave them to
Sakron, a farmer with a stretch of land bordering my own. Sakron is a
friend with a large healthy family. He is honest and he works as if he is

honored to be alive. But Sakron is not family. These fields know family only, and my son is the only living person with the ability to answer their centuries-old longing to be held by the same honoring family.

No, my son will never come home to the fields. He has changed. He has been adopted by a new world that I suspect bad things from and, therefore, do not care to consider. He has often asked me to leave these fields, to go to Canada with him, to live in a house with good running water and a refrigerator. But I know I cannot go. What are these things to me? What will my hardened hands do without the bent plough and the rippling sheaves in the fields? These things have been me since before I was born. They are in the blood of my father and my mother and all those who came before them. They are the agents of the most vital union there is—the union between man and life itself. They are bonding and holy things, and I breathe them and smell them and pay daily homage to them with my nurturing toil and undying love and reverence. If I were to leave them and go to a house in Canada I would surely die of a spiritual torpor begotten by the separation.

I am not a true Buddhist. A true Buddhist is generous and forbearing. He is meritorious in every way, and he is not judgmental of the ways of others. Meditating on his own affairs is enough of a job, for sure, but I cannot keep only to myself. I am weak in that respect, and I am hurt by what will come of my son and my fields when one does not return to the other to disallow the death of the way.

Yes, it is true that I have endured drought and my fields have been stricken by pestilence. I have fallen weak from poverty and have fought for breath when illness came. There have been bandits who steal from our fields, and there have been deadly storms that prey on man and harvest alike. But these things are no arguments against my way. They are the natural relatives of contentment and happiness. I have climbed to the apex of life through my associations with hardship and death. Many have died throughout the years, and many more will die. I have seen it all. And I know that I too will die, maybe in one month, or maybe not for several years yet. But death itself is not so bad. In my own way I have come to love it, for it is an indifferent and honest thing. It makes the good times precious and it employs the hard times as warnings to be thankful for the privilege of life. Without hardship and death, how does one come to understand happiness and life?

I know my new-world son believes me to be ignorant. My refusal to consider his world worries him, and he often gives in to impetuosity and angry self-righteousness, and he tries to persuade me of my world's uselessness. But I am not ignorant. My restfulness is a component of peace—the peace that comes with the recognition of purpose. And what is this purpose? he has asked. To that I have replied that I have lived a good life. I have been charitable when I have been able, I have worked hard without complaint, and I have neither deceived nor felt angry when deceived. I have harmed nobody, neither with action nor with thought. My purpose has been to pay honest, continual respect to the life that I have been given to live.

But my son does not adopt my thinking. Perhaps he believes me to be naive in my simplicity. I do not know. But in his eyes I see a fear—a fear of everything that is me and my village—and I must confess, the essence of this fear I do not understand. I know only that it kills a little of me every time I see it. My son, my own son who once found so much joy and love in simplicity, now forgets. It seems his early village life has been jolted from his memory as if it no longer has function. He forgets the riding of the buffaloes through the flooded fields of the rainy season. He forgets the care-free swimming and the simple games in the water-holes with his friends. And there is no pleasure in his eyes at the sight of bounding and skipping children, happily hunting mud-crabs and minnows, just as he used to do. His face shows no expression.

Or perhaps he chooses to forget. When he visits, he quietly refuses to see his childhood friends who still live in this and other villages, and he is loath to remove his heavy, useless shoes, even in the prickly swelter of the hot season. His communications with the villagers are cordial but distant. He replaces earnestness with politeness. I am often presented with the thought that this reticent young man was not born of the home, but is a visitor, passing through on his way to a different, more comforting destination. Something has died in my son. I cannot help but wonder that I have died in my son.

I am old and seldom leave my village, but I have ears and I hear stories of Bangkok. I hear that it too is forgetting. People, young and old, are convinced of a happiness to be discovered in change: to do different is to do better. The word fashion and style have a meaning, a new importance in Bangkok. Bluejeans sell at high prices. Children no longer care for mangosteen and papaya: they would rather part with their last baht and sip on a Coca Cola than take from the rich earth what is there to be taken. The markets are becoming centers for the old people only. Youngsters take to the air-conditioned shopping malls and pay for their goods, not with barter and bargain, but with money. Goods are now purchased according to a price stamped upon them and all dealings are fast and impersonal. In the village, commerce is more a matter of mutual favour and help. There is no insurance man or bank. There is only the performing of function, be it the buying of cloth, the selling of rice or the trading of help. You know the curry vendor and waver, and you do not suspect them of deceit. Deception and cheating are the methods of only a very few who travel from village to village, selling what they can then moving on. But here, a deal is made out of need, rarely want, and is therefore a matter of understood importance. I do not know how much longer this will be. I am told that Bangkok is forgetting more and more each day.

But my son loves Bangkok. It is where he found the path to his new world. He went to a school there, and through his good work he won a chance to study in a better school in Canada. The path took him, in six short years, from a slum in Bangkok to what he calls a clean and beautiful city in Canada. Will this path continue to grow? Will it one day reach out to find this village and replace generations of simple method with ill-directed passion and the love of excess?

But I do not wish to be judgmental. I speak only in respect to my own feelings and therefore do not have much to say. Who am I, a farmer who has only left his fields and village twice to visit friends in Bangkok, to disapprove of change when there are so many who purport to become champions of it? I should simply say that I do not care to consider it and leave it at that, for you do not eat a meal when your stomach is full. But why are more stomachs not full in the way that mine is?

As answer, (and I do not mean to speak generally, but of what may only be a few) I think some seek to overfill. Friends convince me of a new selfishness. These friends, like me, are farmers with old notions of simplicity and fate. But they possess large stretches of beautiful land, too vast and too demanding to be satisfied with the nurturing of a single pair of hands. Yearly, these farmers rely on helping hands, young men without fields and homes, who give their bodies to harvesting for money. In past years, the farmers allotted forty to fifty baht to each hand for a day's work. But I am told now that the hands are asking as much as eighty or ninety baht for the same work. There is a new solidarity among the hands, a belief that their increasing demands, when

tempered by collective response, will be met. And what can the farmers do? Their fields must be harvested, so they meet the new demands and watch as the little money that they make from their crops is devoured by the hands.

As reason for their demands the hands mention the danger created by the Khmer shells that occasionally soar over the border and land in the fields. But this is no excuse. No person is ever killed by these shells, for they are infrequent occurrences, usually coming at night when all are in the village, resting and drinking Hong Yok after a day of harvesting. But should such a danger exist, is it not obvious that money is no compensation for a thing so precious as life? Of course, it is in our nature to avert death with every trace of purpose we can find. But to make profit from it, to equate it with anything in this world and say that thing is of equal power, is to teach yourself a grave and sad dishonesty. I wonder if these hands have ever felt the death that comes with war. If they have, I should gladly listen to their reasoning, for at present, I am unable to understand them except as ill-grounded selfishnesses.

It has been my hope to speak honestly about my son and his visit to the village, but I fear I have fallen into another subject. It is not my right to be portentous, and I apologize if I have overstepped my boundary. My duty is to meditate upon my own life, and to be good and charitable when I am able. But I am not a true Buddhist, and when it comes to my son and my fields and the only way I have ever known I become impassioned beyond the point at which it is perhaps best to speak. I do not mean to imply that, like the hands, my son is selfish. No, his love for change formed honestly, deep within his heart, and there is no horrible motive in his method. I know that within his own definition of the word my son is honest.

But honesty does not always lead to happiness, and because his is a devoted agent of change, my son is a footsoldier in a surging, reckless legion. His change does not know when enough is enough. It is like a foolish dog whose greatest felicity does not come from the simple recognition of the beat of its own heart. Most dogs will eat, procreate and defend themselves when threatened; all as methods to this end. But a foolish dog will also seek ease when he has mastered the skill of keeping alive. Instead of finding and tracking his food in the fields he will overturn baskets of fish, searching for easy food; he will enter huts, smelling the boiling buffalo meat or the salted fish, and he will inevitably be sent reeling by the kiss of a broom or the thrust of a foot. Some dogs do not learn from their folly, for there is no questioning of method. There is only an unquenchable desire for more. Change is no benign thing when one does not recognize one's own comfort.

My son has also learned silence. He ignores the pains of his heart and locks them in a dark corner of his mind. He does not bring his wife to Thailand with him, fearing she has not developed a vision for this way of life. Rather than show her his early home, he leaves her in his house in Canada where she must find peace from something she can only guess about. My son is embarrassed and apologetic when he tells me this. I hear in his voice a soft wistfulness, a suppressed shame of his wife and of himself. Yet he accepts this shame rather than admit to the truth that if she does not possess the eyes for this village, this world that gave birth to him, she does not possess the eyes for him. His birthplace is his earliest root, and try as she might, she cannot forget it and she must not allow him to deny it, for without it they will not continue to grow, and they will never be able to spread new roots in the new-world direction that change is inspiring them to. The heart cannot guide when you do not trust its strength.

I do not communicate these thoughts to my son because I know he infers them from my manner anyway. Words do not need to be spoken when actions tell a truth. He rarely looks me in the eyes to gladly show me something of himself. There is always the contrived indifference, the fear, and the weak, transparent purpose of concealing it. But perhaps he sees more of my concerns than I think he does, and perhaps he feels the tug of a predicament that he does not yet understand. I cannot say for sure that he does not secretly long for a return to this village, to the impossible, long ago method that has escaped him and seems too far away to return to. Perhaps he too, in his own way, fears the death of the way.

Sometimes when the day is long and hard and my breath is short, I look up from my calloused hands and the earth-ground plough that they grip, and I see my son standing brilliantly and proudly before me, the sunlight resting on his baked and muddied skin. I rejoice that he has returned to his home to claim the fields that are in his blood and the way that is hidden deep in his heart. But the vision inexorably fades and I am alone and the fields are alone with me. And my son is in his new world, learning to forget that the ghosts of these fields are lamenting the coming end of the only way they have ever known.

A Moving Day

BY

SUSAN

NUNES

~

cross the street, the bulldozer roars to life. Distracted, my mother looks up from the pile of embroidered linen that she has been sorting. She is seventy, tiny and fragile, the flesh burned off her shrinking frame. Her hair is gray now—she had never dyed it—and she wears it cut close to her head with the nape shaved. Her natural hairline would have been better suited to the kimono worn by women of her mother's generation. She still has a beautiful neck. In recent years she has taken a liking to jeans, cotton smocks, baggy sweaters, and running shoes. When I was a child she wouldn't have been caught dead without her nylons.

Her hands, now large-jointed with arthritis, return to the pile of linen. Her movements always had a no-nonsense quality and ever since I was a child, I have been wary of her energy because it was so often driven by suppressed anger. Now she is making two stacks, the larger one for us, the smaller for her to keep. There is a finality in the way she places things in the larger pile, as if to say that's *it*. For her, it's all over, all over but this last accounting. She does not look forward to what is coming. Strangers. Schedules. The regulated activities of those considered too old to regulate themselves. But at least, at the *very* least, she'll not be a burden. She sorts through the possessions of a lifetime, she and her three daughters. It's time she passed most of this on. Dreams are lumber. She can't *wait* to be rid of them.

My two sisters and I present a contrast. There is nothing purposeful or systematic about the way we move. In fact, we don't know where we're going. We know there is a message in all this activity, but we don't know what it is. Still, we search for it in the odd carton, between layers of tissue paper and silk. We open drawers, peer into the recesses of cupboards, rummage through the depths of closets. What a lot of stuff! We lift, untuck, unwrap, and set aside. The message is there, we know. But what is it? Perhaps if we knew, then we wouldn't have to puzzle out our mother's righteous determination to shed the past.

There is a photograph of my mother taken on the porch of my grandparents' house when she was in her twenties. She is wearing a floral print dress with a square, lace-edged collar and a graceful skirt that shows off her slim body. Her shoulder-length hair has been permed. It is dark and thick and worn parted on the side to fall over her right cheek. She is very fair; "one pound powder," her friends called her. She is smiling almost reluctantly, as if she meant to appear serious but the photographer has said something amusing. One arm rests lightly on the railing, the other, which is at her side, holds a handkerchief. They were her special pleasures, handkerchiefs of hand-embroidered linen as fine as rice paper. Most were gifts (she used to say that when she was a girl, people gave one another little things—a handkerchief, a pincushion, pencils, hair ribbons), and she washed and starched them by hand, ironed them, taking care with the rolled hems, and stored them in a silk bag from Japan.

There is something expectant in her stance, as if she were waiting for something to happen. She says, your father took this photograph in 1940, before we were married. She lowers her voice confidentially and adds, now he cannot remember taking it. My father sits on the balcony, an open book on his lap, peacefully smoking his pipe. The bulldozer tears into the foundations of the Kitamura house. What about this? My youngest sister has found a fishing boat carved of tortoise shell.

Hold it in your hand and look at it. Every plank on the hull is visible. Run your fingers along the sides, you can feel the joints. The two masts, about six inches high, are from the darkest part of the shell. I broke one of the sails many years ago. The remaining one is quite remarkable, so thin that the light comes through it in places. It is delicately ribbed to give the effect of canvas pushed gently by the wind.

My mother reaches for a sheet of tissue paper and takes the boat from my sister. She says, it was a gift from Mr. Oizumi. He bought it from an artisan in Kamakura.

Stories cling to the thing, haunt it like unrestful spirits. They are part of the object. They have been there since we were children, fascinated with her possessions. In 1932, Mr. Oizumi visits Japan. He crosses the Pacific by steamer, and when he arrives he is hosted by relatives eager to hear of his good fortune. But Mr. Oizumi soon tires of their questions. He wants to see what has become of the country. It will be arranged, he is told. Mr. Oizumi is a meticulous man. Maps are his passion. A trail of neat X's marks the steps of his journey. On his map of China, he notes each military outpost in Manchuria and appends a brief description of what he sees. Notes invade the margins, march over the blank spaces. The characters are written in a beautiful hand, precise, disciplined, orderly. Eventually, their trail leads to the back of the map. After Pearl Harbor, however, Mr. Oizumi is forced to burn his entire collection. The U.S. Army has decreed that enemy aliens caught with seditious materials will be arrested. He does it secretly in the shed behind his home, his wife standing guard. They scatter the ashes in the garden among the pumpkin vines.

My grandfather's library does not escape the flames either. After the army requisitions the Japanese school for wartime headquarters, they give my mother's parents twenty-four hours to vacate the premises, including the boarding house where they lived with about twenty students from the plantation camps outside Hilo. There is no time to save the books. Her father decides to nail wooden planks over the shelves that line the classrooms. After the army moves in, they rip open the planks, confiscate the books, and store them in the basement of the post office. Later, the authorities burn everything. Histories, children's stories, primers, biographies, language texts, everything, even a set of Encyclopaedia Brittanica. My grandfather is shipped to Oahu and imprisoned on Sand Island. A few months later, he is released after three prominent Caucasians vouch for his character. It is a humiliation he doesn't speak of, ever.

All of this was part of the boat. After I broke the sail, she gathered the pieces and said, I'm not sure we can fix this. It was not a toy. Why can't you leave my things alone?

For years the broken boat sat on our bookshelf, a reminder of the brutality of the next generation.

Now she wants to give everything away. We have to beg her to keep things. Dishes from Japan, lacquerware, photographs, embroidery, letters. She says, I have no room. You take them, here, *take* them. Take them or I'll get rid of them.

They're piled around her, they fill storage chests, they fall out of open drawers and cupboards. She only wants to keep a few things— her books, some photographs, three carved wooden figures from Korea that belonged to her father, a few of her mother's dishes, perhaps one futon.

My sister holds a porcelain teapot by its bamboo handle. Four white cranes edged in black and gold fly around it. She asks, Mama, can't you hang on to this? If you keep it, I can borrow it later.

My mother shakes her head. She is adamant. And what would I do with it? I don't want any of this. Really.

My sister turns to me. She sighs. The situation is hopeless. You take it, she says. It'll only get broken at my place. The kids.

It had begun slowly, this shedding of the past, a plate here, a dish there, a handkerchief, a doily, a teacup, a few photographs, one of my grandfather's block prints. Nothing big. But then the odd gesture became a pattern; it got so we never left the house empty-handed. At first we were amused. After all, when we were children she had to fend us off her things. Threaten. We were always *at* them. She had made each one so ripe with memories that we found them impossible to resist. We snuck them outside, showed them to our friends, told and retold the stories. They bear the scars of all this handling, even her most personal possessions. A chip here, a crack there. Casualties. Like the music box her brother brought home from Italy after the war. It played a Brahms lullaby. First we broke the spring, then we lost the winding key, and for years it sat mutely on her dresser.

She would say again and again, it's impossible to keep anything nice with you children. And we'd retreat, wounded, for a while. The problem with children is they can wipe out your history. It's a miracle that anything survives this onslaught.

There's a photograph of my mother standing on the pier in Honolulu in 1932, the year she left Hawaii to attend the University of California.

She's loaded to the ears with leis. She's wearing a fedora pulled smartly to the side. She's not smiling. Of my mother's two years there, my grandmother recalled that she received good grades and never wore a kimono again. My second cousin, with whom my mother stayed when she first arrived, said she was surprisingly sophisticated—she liked hats. My mother said that she was homesick. Her favorite class was biology and she entertained ambitions of becoming a scientist. Her father, however, wanted her to become a teacher, and his wishes prevailed, even though he would not have forced them upon her. She was a dutiful daughter.

During her second year, she lived near campus with a mathematics professor and his wife. In exchange for room and board she cleaned house, ironed, and helped prepare meals. One of the things that survives from this period is a black composition book entitled *Recipes of California*. As a child, I read it like a book of mysteries for clues to a life which seemed both alien and familiar. Some entries she had copied by hand; others she cut out of magazines and pasted on the page, sometimes with a picture or drawing. The margins contained her cryptic comments: "Saturday bridge club," "From Mary G. Do not give away," underlined, "chopped suet by hand, wretched task, bed at 2 A.M., exhausted." I remember looking up "artichoke" in the dictionary and asking Mr. Okinaga, the vegetable vendor, if he had any edible thistles. I never ate one until I was sixteen.

That book holds part of the answer to why our family rituals didn't fit the recognized norm of either our relatives or the larger community in which we grew up. At home, we ate in fear of the glass of spilled milk, the stray elbow on the table, the boarding house reach. At my grandparents', we slurped our *chasuke*. We wore tailored dresses, white cotton pinafores, and Buster Brown shoes with white socks; however, what we longed for were the lacy, ornate dresses in the National Dollar Store that the Puerto Rican girls wore to church on Sunday. For six years, I marched to Japanese language school after my regular classes; however, we spoke only English at home. We talked too loudly and all at once, which mortified my mother, but she was always complaining about Japanese indirectness. I know that she smarted under a system in which the older son is the center of the familial universe, but at thirteen I had a fit of jealous rage over her fawning attention to our only male cousin.

My sister has found a photograph of my mother, a round-faced and serious twelve or thirteen, dressed in a kimono and seated, on her knees, on the *tatami* floor. She is playing the *koto.* According to my mother, girls were expected to learn this difficult stringed instrument because it was thought to teach discipline. Of course, everything Japanese was a lesson in discipline—flower arranging, calligraphy, judo, brush painting, embroidery, everything. One summer my sister and I had to take *ikebana,* the art of flower arrangement, at Grandfather's school. The course was taught by Mrs. Oshima, a diminutive, soft-spoken, terrifying woman, and my supplies were provided by my grandmother, whose tastes ran to the oversized. I remember little of that class and its principles. What I remember most clearly is having to walk home carrying, in a delicate balancing act, one of our creations, which, more often than not, towered above our heads.

How do we choose among what we experience, what we are taught, what we run into by chance, or what is forced upon us? What is the principle of selection? My sisters and I are not bound by any of our mother's obligations, nor do we follow the rituals that seemed so important. My sister once asked, do you realize that when she's gone that's *it*? She was talking about how to make sushi, but it was a profound question nonetheless.

I remember, after we moved to Honolulu and my mother stopped teaching and began working long hours in administration, she was less vigilant about the many little things that once consumed her attention. While we didn't exactly slide into savagery, we economized in more ways than one. She would often say, there's simply no time anymore to do things right.

I didn't understand then why she looked so sad when she said it, but somehow I knew the comment applied to us. It would be terrible if centuries of culture are lost simply because there is not time.

Still, I don't understand why we carry out this fruitless search. Whatever it is we are looking for, we're not going to find it. My sister tries to lift a box filled with record albums, old seventy-eights, gives up, and sets it down again. My mothers says, there are people who collect these things. Imagine.

Right, just imagine.

I think about my mother bathing me and singing, "The snow is snowing, the wind is blowing, but I will weather the storm." And I think of her story of the village boy carried by the Tengu on a fantastic flight over the cities of Japan, but who returns to a disbelieving and resistant family. So much for questions which have no answers, why we look among objects for meanings which have somehow escaped us in the growing up and growing old.

However, my mother is a determined woman. She will take nothing with her if she can help it. It is all ours. And on the balcony my father knocks the ashes of his pipe into a porcelain ashtray, and the bulldozer is finally silent.

My Grandparents

BY

RUTH

FAINLIGHT

Museums serve as my grandparents' house.
They are my heritage—but Europe's spoils,
Curios from furthest isles,
Barely compensate the fact
That all were dead before I was alive.

Through these high, dust-free halls, where
Temperature, humidity, access,
Are regulated, I walk at ease.
It is my family's house, and I
Safe and protected as a favoured child.

Variety does not exhaust me.
Each object witness to its own
Survival. The work endures beyond
Its history. Such proof supports me.
I do not tire of family treasures.

Because no one remembers who they were,
Obscure existences of which I am
The final product, I merit
Exhibition here, the museum's prize,
Memorial to their legend.

A
Poet's
Story

~

BY

JULIO

HENRIQUEZ

ne day, a day just like any other, my younger brother died. In a very confusing investigation, all we could find out was that the vehicle in which he was driving with a friend was hit by a heavy truck.

One day, a day just like any other, my niece Carminda disappeared. She was captured by men in plain clothes. Her father was never to see her again.

One day, a day just like any other, Jaime, a poet, was arrested in town. Later on, we saw his body, savagely tortured and mutilated. Just three days before his assassination we had been planning a literary page for a local newspaper, exploring the theme of cultural identity.

Another day, another day like any other, I got off the bus at the university and was told that a death threat had been received by our newspaper. On the Tuesday, they caught me. There were six of them. The fat one, who carried a machine gun, blindfolded me and forced me into a canvas sack. "We know that you are the poets of the guerrillas," said their chief. "You had better talk. Don't be stupid. You don't know where you are, and we are not the police."

We went up a big staircase and they locked me in a room with the air conditioner turned on full. It was cold, cold, cold. I spent two days there, my jaws shivering. They didn't remove the blindfold or the canvas sack.

What must my wife be thinking? And the kids? The boy will not despair. The girl is small, but she already understands. And my mother is strong, really strong. You are intelligent and well educated, I thought to myself. Why do you get involved in this dangerous stuff? You see, none of your friends are asking about you. They haven't said anything. But I won't give in. I swear I won't give in, they won't make me give in.

Seven days went by before I was released. I felt lucky and happy. Happy because they didn't make me give in. Because they could have killed me and I didn't care. Because they didn't get a single name out of me. I didn't betray my brothers, my friends, my buddies, my family. Or my conscience.

But the death threats continued. Many friends disappeared or were kidnapped. For many days I could not step foot near my house. Someone working with the Archbishop said to me, "the country needs you here, but the people need you alive. Wherever you get to, wherever you happen to be, continue to write." After our conversation I presented myself to the Mexican embassy and asked for a visa so that I could leave the country.

On 15 October 1985 I left. I did not have the heart to say goodbye to my family. I walked through the airport without turning my back, because I knew that my wife, my children and my parents were watching me. I had not cried for a long time, and I did not want to cry then.

Mexico is a country which I carry in my heart. The Mexican people are just like mine. Our blood is the same. It is a beautiful country where I got together with old friends who I had not seen for a long time. That's where the world began to hug me, and where I had the chance to make contact with UNHCR [United Nations High Commissioner for Refugees].

It took me six months to get my papers ready for Canada. I arrived in Toronto anxious and uncertain, experiencing culture shock. I was overwhelmed with doubt, not knowing how long it would be before I could enjoy the indispensable warmth of my family. I knew from letters I received that my son had graduated from music school and was working as a member of the union of artists and cultural workers. And my youngest daughter was still in dance school, a member of the children's national ballet company.

I was attending an English course one morning when I was called to the office. They gave me the news that my wife and children were arriving the following month. When they came, we gathered in the lobby of a hotel. I felt happy but sad, lucky but sad, rejuvenated but sad. Because that was when I realized the struggle they had to go through. They had suffered and been pushed around, and now I had pulled them away, chopping them off from their roots.

To re-establish a lifestyle in exile is not easy, but many people in Hamilton showed us generosity and kindness. I found a job in a factory, where I am treated well. After completing a training course, my wife managed to start working at a university cafeteria. She is studying English and becoming more and more confident. Perhaps the best thing has been the benefits available to my children. My son is a member of two local orchestras, and he is working on a local project which provides information about the culture of my homeland. My daughter is still learning to dance, and is thinking about joining the university art school.

But will I ever be back home, walking through the valleys and hills of my small Central American country? I don't know. I only hope that life continues to nourish me with an understanding of the citizens of the world. Because I have learned that when you have only one breath left, when you have one last penny in your pocket, your solitude and the love you have inside you are enough to enable you to start again.

Sisters

BY

SIBANI

RAYCHAUDHURI

When I was small, we always thought that everything would be all right once we came to England. Shirin, my older sister, had shown me the black dot on the map, which marked the city where our father lived, and had told me we would go there one day. I thought of England as a very special place that would bring an end to all our problems.

In those days we lived in a little village in Bangladesh where my father's family had lived for as long as anyone could remember. Amma did what she could to keep us clean and fed, but it was Shirin who always played with me and looked after me. Our main excitement, each week, came on those days when we saw the postman walking towards our house with his heavy shoulder bag. If there were any letters from Abba he would shout out *"Chitti!" "Chitti!"*and we would run into the lane to meet him. When the letters arrived, Shirin was the one who had to read them to us and Amma was always impatient. "Is there any news?" she would ask, before Shirin had even had time to cut open the aerogramme. We knew what her question meant without her having to complete it. We had been waiting for over five years to get permits to go to England. I had seen Abba only twice in that time. He could not come more often than that: he was saving money for our fares. I hardly knew him, but I missed him a lot. I kept all his letters in a biscuit-tin, which he had brought for us on his first visit. The last time he came, he had arrived home with a huge case full of new clothes for us children, nylon saris for Amma, and a small radio: Shirin and I had fought over it until Amma had threatened to give it away unless we learned to share it.

We had good meals every day for a month, while Abba was there, and neighbours and relations crowded our house all hours of the day. He sat on the veranda with them. They talked and talked, while Amma and Shirin made endless cups of tea for them. When everyone had gone in the late evening, he told us stories: stories of his childhood and stories of England, that faraway place where, one day, we would all live.

Every month Abba sent money from London. When the money arrived, Shirin went with Amma to the post office to collect it. Each moment at the post office was a torture for her; she had to prove who she was, knowing that if anything went wrong she would not be given the money. Once when they were getting ready to go to the post office, I started to get dressed as well. Shirin was puzzled: "Why are you putting on a clean dress at midday?"

"I am taking Amma to the post office."

She smiled. "Do you know what to do there?"

"I could find out. Why can't I go for once? You go with Amma every time." Shirin loved me dearly, so she pleaded with Amma: "Let Nazia take you this time. I'll tell her what to do."

But Amma refused to listen: "Nazia can stay and look after Kamal," she said. I knew there was no point in arguing.

When Amma and Shirin came back, I sulked and ignored them. They didn't take any notice of me either. They were busy unpacking the groceries: sugar, tea, lentils, salt, spices. I sat on the edge of the veranda watching Kamal as he played on the baked red earth of the courtyard. He was making a series of small holes in the ground for a game of marbles. Suddenly I felt a hand over my eyes: "Nazia, guess what we have got for you?"

I turned round and there was Shirin standing behind me with a length of silk ribbon. This was a great luxury in those days. I knew Amma had little money to spend on extras. The money Abba sent home was just enough for bare necessities. Whenever we asked for something, Amma would chide us: "That poor man sweating over a hot oven day and night: what do you know of life?" All I knew was that Abba had to work very hard over there in England to keep us going in Jagganathpure. And I looked forward to the day when we would be allowed to join him.

Every monsoon our village became water-logged for days: the narrow path leading to our house from the road disappeared; the courtyard flooded—the water rose to the edge of the veranda. We lay awake at night, unable to sleep, listening to the storm. The wind rattled the corrugated-tin roof, and the heavy rain hissed through the leaves of the trees behind the house. I remember one year in particular. Shirin and I huddled together on the bed; while Kamal lay asleep on Amma's lap. We heard the frogs croaking outside in the darkness. Then we heard: *Splash! Splash! Splash!* Something came closer and closer, and then disappeared. We spent the rest of the night in fear. Shirin held me tight: "Don't be frightened, Nazia. Look out of the window—see the lights in the schoolhouse."

Amma heard our whispering: "If it goes on like this, we may have to move to the schoolhouse ourselves."

Next morning, we could see the damage. Our kitchen, a small clay hut in the corner of the courtyard, had lost its roof. Our pots and pans had washed out under the half-door and now floated in the courtyard along with vegetable-peel, burnt wood from the fire, and the body of a drowned cat. The cat had belonged to our neighbours. Kamal and I wept when we saw it. Our small vegetable patch was gone and the top of our coconut tree had been broken by the wind. Shirin wiped Kamal's eyes and put her arms round our shoulders. "Don't worry," she said, "we won't be here next year." And we thought how everything would be all right when we went to London. Dry floors, dry clothes, dry beds. "Have you been to London?" asked Kamal. Shirin laughed and shook her head: "Abba has been trying to get us over there since before you were born!" Kamal counted the years on his fingers—five, six, seven!

Then came the day when the postman's shout *"chitti"* brought a registered letter from England. Shirin had to sign a form to receive the letter, while Amma stood beside her anxiously. Kamal and I were very excited and asked hundreds of questions: "Are we going then?" "Has he sent the plane tickets?" We didn't stop until Amma told us off.

"Don't get too excited—there is still a lot to be sorted out." She had become superstitious in those days; she thought that things would not come true if we expected them too confidently.

The letter, however, brought good news, and it seemed that we were on our way, at last. Abba had asked an old friend in the village to help with arrangements at our end. Mr. Ahmed had already done this for another family. "They are living in London now," he told us, a few days later, when he dropped in to see Amma, "on the seventh floor of a very nice, tall block of flats." He searched for something in his pocket. "They have sent me a colour photo." But he could not find the photo. He shook his head and looked thoughtful for a moment. "It's a lot of work. It takes time." Even so, none of us had any idea just how much work it would take before we could board the plane. First we had to go to an office in Sylhet Town. We started our journey very early, before the sun rose. We were dressed up for the occasion. Amma and Shirin wore the nylon saris Abba had brought on his last visit. Amma was nervous. She kept on checking everything again and again. "Have you taken the water-bottle?" she asked Shirin. "Have you got the food safe?" she asked me. Then, when Mr. Ahmed arrived, Amma checked everything once more. We crossed the river by the narrow, bamboo bridge. Then Mr. Ahmed led us across the cracked earth of the dried-out fields towards the road, where we waited for the bus. It was a long and difficult journey: after the bus-ride came a ferry, then another bus-ride and another ferry. It was the first time that Kamal and I had been out of our village.

And, in the end, this first journey came to nothing. When we arrived at the office, a wooden sign was hanging on the door: "Closed." We sat down on the steps, saddened and exhausted. We had had to stand most of the way, and the day had been very hot. Mr. Ahmed sighed. "Just our luck! If that second bus hadn't stopped to collect sacks of rice from the market, we would have been here in time." Shirin got to her feet and tucked the water-bottle under her arm: "Let's go!" she said quietly. "We don't want to spend the night here."

After that first fruitless trip, we made numerous journeys to Sylhet, and, on one occasion, two Englishmen turned up in our village and asked our neighbours some searching questions about us. Amma had felt very tense and worried in case our neighbours made any mistakes—or left anything out. But Kamal and I lived in a state of euphoria. We often skipped going to school. Any day now we would be leaving for England, and, in the meantime, we accompanied Amma wherever she went.

It was nearly two years before we were granted our final interview in Dhaka. A trip to Dhaka was even more of an adventure than a trip to Sylhet Town. We arrived in Dhaka the night before our interview. Mr. Ahmed had arranged for us to stay with his sister, whose husband was the manager of a cotton mill. We spent the night in their guest room, which had paintings on the walls, a brass pot with a plant on a small table in the corner of the room and a chiming clock on the bookshelf. Months later, I saw a clock like that in a shop window in London. Next day, at eight in the morning, we were waiting to go into the High Commission. About twenty women, young and old, all dressed in their best saris, were sitting on the wooden benches. Amma and Shirin joined them. Kamal and I sat on the floor with the other children. A man was standing at a desk: he called out the names one by one. It was midday before Amma's name was called. We followed her into the room.

We entered a large room and stood like convicts in front of a British officer, seated on a high-backed wooden chair. There was a huge shiny table between us and him. A Bangladeshi man was waiting by the officer to interpret for him. The officer asked Amma question after question about our family until there was nothing else she could tell him. Then he turned to Shirin and interrogated her as well: "What was your grandfather's name?' he asked. "His father's name? His mother's name?" Shirin had never met our great-grandparents, but she had been warned to expect such questions. The officer turned back to our mother. He seemed to be worried that Shirin did not have a birth-certificate. Amma was lost for words; she did not know what to say. She tried to explain: Shirin was born at home, and nobody had thought to have her birth recorded. It was not a custom then; everyone knew when she was born. The officer made some notes on the papers in front of him and then took the papers away to another room. We waited in silence, in suspense. Those few minutes seemed to stretch endlessly. When at last he came back, the officer announced the decision.

"We have decided to allow you and your two younger children entrance to Britain to join your husband, but we have decided not to grant entry to Shirin Bibi. She is eighteen and is, therefore, too old to go as your dependant." I could not believe what I heard. Shirin wouldn't be coming with us. After all those years of waiting, Shirin was now "too old."

All the way back home, Amma was wiping her eyes on the end of her sari. Shirin didn't say a word, she only held Amma's hand tightly. Amma tried to reassure her: "Once your Abba hears this, he'll do all he can to bring you to join us. He must know a lot of people there." Shirin just nodded.

All our excitement about going to England had dried up at the office in Dhaka, but here was no time to brood, there was so much to do before we left. We had to arrange for Shirin to stay with our Khala, Amma's sister, in the next village. Our Khala came to collect her a few days before we caught the plane. Since we had got back from Dhaka, I had followed Shirin like a shadow wherever she went. We were going to be separated for the first time in our lives. I couldn't imagine what life would be like without her, and I didn't want to lose a minute of her company in those last days. On the day she left, when I helped her pack her clothes, I could not speak to her—my voice was choked with tears. When we finished Shirin gave me a big hug, and I handed over to her our prized possession, the small radio. Shirin could no longer keep herself under control, and the tears flooded her eyes. We sat holding each other on the floor of the little room we shared. Then Khala came into the room with Amma and said, "Come on! We must hurry or we'll miss the bus." I went to the bus-stop with them—and I stood, watching and crying, long after the bus had disappeared over the dusty horizon.

The dream had come true. We had come to England, but we had left Shirin behind us. Six families lived in the house Abba brought us to. We had one small room with two beds and a pile of blankets and pillows. We had to take turns for cooking, washing and taking baths. We had to dry our clothes in the room which smelt of damp and paraffin. There were always petty rows when people got in each other's way. Amma became nervous and frightened, and stayed in our room all the time. Abba was never there: he left for work early and came home late.

One day Kamal asked, "Where can we play? Can we play in the street?"

Abba suddenly became anxious: "Don't you two go out of this house on your own. I'll get you some games— you can play in the room."

But Abba didn't buy us any games, and Kamal and I played on the stairs with the other children of the house.

Kamal and I were sent to different schools. Kamal's was a small primary school across the road, but mine was ten minutes' walk from home. It was big and confusing, a secondary school crowded with children. There were fights in the playground and fights in the corridors. There was lots of jostling and pushing whenever we changed rooms. On my first day I was terrified. At break I was waiting by the water-fountain to have a drink. I was standing at the edge of a crowd of boys and girls. Some were drinking; others were splashing water on us.

"Are you new here?" someone asked me in Sylheti. I nodded. Then she pulled me by the arm towards the fountain. "You won't get anywhere here, if you wait quietly," she said, "you have to fight for everything."

This was Rukeya, and this was the start of our friendship. As we walked back across the playground, she said: "I'll take care of you until you get used to things."

After that, I stayed with Rukeya. We spoke the same language, and, besides, most of the other girls didn't want to mix with us.

One day, when I was running down the corridor with Rukeya between lessons, two of them grabbed hold of us. They pulled our hair for no reason at all, and we punched them to make them let go. A small group of girls gathered to watch. They chanted in chorus: "Fight! Fight!"

The noise brought out the headmistress and we were marched into her room. Then Maxine and Fay told their versions of the story. Rukeya and I tried to tell ours. Ms. Butler didn't seem to listen to us. She went on and on about our bad behaviour: "This is disgraceful! Punching and scratching each other!" Rukeya and I kept our heads down, but we couldn't help noticing what Maxine was doing. Whenever Ms. Butler looked in our direction and away from the other two, Maxine made a large pink bubble with the gum in her mouth. Fay, like us, was struggling not to laugh. Ms. Butler suddenly shouted at her, "You can wipe that silly grin off your face!"

Fay drew the back of her hand across her mouth, and the three of us giggled. That made Ms. Butler furious. "You are all on report," she said. She handed a pink card to each of us and told us we had to ask our teachers to sign it at the end of each lesson.

We walked quietly out of her room, but as soon as we had turned the corner at the end of the corridor, we roared with laughter. Maxine nudged me: "Have one." She held out a packet of bubble-gum. I hesitated for a moment, but Rukeya said, "Go on—take one!"

On Saturdays, there were stalls on the pavement by the tube-station, and a market in the open space behind the shops. Rukeya and I went round the market to do the weekly shopping for our families. We gave Kamal the job of carrying the bags. As yet Kamal didn't have any friends to go out with and he didn't protest too much. One Saturday, when Kamal couldn't come with us, we went to "City Girls" in the High Street. The shop was darker than any other we'd ever been to, and music was playing loudly. We moved shyly between the racks, feeling the smoothness of the long silky dresses. Quite unexpectedly, I felt a touch on my shoulder. Someone whispered: "Hi! I didn't know you shopped here." I turned round to see who it was. It was Maxine. We were astonished and impressed! She was doing the same as us, only it was our first time at City Girls whereas she had clearly been there often before. Then she winked at us and led us down a spiral staircase to the basement. Here were the changing rooms—a series of cubicles with drop curtains. We exchanged glances: we knew exactly what we were going to do now. Rukeya and Maxine returned to the ground floor, and each took a long dress from the racks. Maxine's was black and Rukeya's was deep purple. I tried on a pair of tight leather trousers and a short jacket spotted like leopard-skin. We admired ourselves in the mirror and praised each other's choices with mischievous smiles. Rukeya and I knew that our parents would never let us dress up as "City Girls."

Every Sunday morning, Rukeya and I went to the community centre for our Bengali lessons. It was more of a club than a class. Besides our language lessons, we were allowed to listen to songs and to watch Bengali films on video. There was also a small library, which kept Bengali books, newspapers and magazines. We used to get news of Bangladesh from the centre. A few days before our school holiday ended, Rukeya called to see me and we went out together to the centre. When we entered the hallway, a crowd was standing there. They were looking at the front page of a newspaper and talking anxiously. We wanted to know what was going on. They showed us the front page: there were photographs of bodies floating in water, people standing on a roof-top. There had been a terrible flood in Bangladesh!

Abba got home late as usual. He had already heard the news about the flood, and he looked very worried. He was restless; he couldn't stay in the room, he went to talk to other people in the house. Then he went out to see someone he knew who lived in the next street. Amma was shaking with fear about what might have happened to Shirin. I couldn't calm her down. When Abba came back he told us what he'd heard: the floods had washed away many villages on the coasts, and the small islands in the Bay of Bengal were under the sea. The electric and telephone lines had been damaged by the storm, and people were cut off from the outside world; only a few helicopters were dropping food and blankets into the damaged areas. Abba tried to assure us: "Your Khala does not live on the coast. I heard that people in other places had taken shelter on higher ground."

Amma wept and kept repeating: "My Shirin must be cold and hungry. How can I eat if she hasn't had a grain of food?" That night, when I went to bed, I sat up and prayed that Shirin was safe.

The summer holidays came to an end and we had still had no definite news of Shirin. Abba gathered from his friends that my Khala and her family were safe, but they had lost their home and nobody knew where they were staying. Friends kept on telling us "No news is good news," but Amma was so depressed that she stayed in bed. I was too upset to go back to school; I stayed at home to look after Amma. Kamal went to school with other boys from the house. Every day after school, Rukeya came to see me. She used to say, "Nazia, please come to school—it'll take your mind off your worries."

One evening when I opened the door for Rukeya, Maxine was standing outside with her. "Can we come in?" Rukeya asked. "Yes," I replied hesitantly. I rushed to tidy up the bed-clothes to make room for them. Amma was puzzled by this unexpected visit. Maxine sat on the bed next to Rukeya and handed me a letter from our form tutor: "We have told Miss Richards about Shirin and your aunt. She has seen the news of the floods on television. We are raising money for flood-relief."

Rukeya broke in: "We'll give the money to the community centre, and they'll make sure it gets where it's needed." Amma wanted to know what was going on, and why they were talking about Shirin and Bangladesh. Rukeya repeated everything in Sylheti. Amma gave thanks to Allah. When Maxine and Rukeya left, they made me promise to come back to school next day.

Next morning I went back to school as promised. On my way there I felt I had lost touch with everything over the last month. I was completely blank when I arrived. The first lesson was a form-period. Miss Richards accepted my absence note and didn't ask me for any further explanation. She suggested that I should take a look at some of the work the other children had done in their project on Bangladesh. The work was displayed on the wall. There was also a box of books from the library. I picked up a Bengali book, *Our Bangladesh*, illustrated with large colour photographs. I sat down in the back of the classroom and started reading. Miss Richards was surprised to find that I could read Bengali; she was curious and asked me if I could translate to her what I had read. For the first time I felt like an expert.

She encouraged me to get involved in the project, and Rukeya and I went round our friends and neighbours to borrow things for display, things to read, things to listen to. They lent us photographs, things made from bamboo, tapes of *Amar Sonar Bangla* and songs by Nazrul. Rukeya borrowed her Amma's silk sari to hang behind our display. We also collected newspaper cuttings and picture-postcards from the community centre. We put a lot of effort into the project. Whatever I did, I felt I was doing it for Shirin.

When the project was completed, our class mounted an exhibition of our work in the foyer of the school. Miss Richards put a collection box in the corner of the exhibition. Boys and girls from other classes hung curiously round the place with interest. Rukeya and I and two other Bangladeshi girls acted as guides. One day, Miss Richards brought a large map of the world to our class. She pinned it up on the wall and asked me to come to the front of the class. "Nazia," she said, "can you show on the map where you were born?"

I soon found Bangladesh and Sylhet, but I found it hard to locate our village. I was sure it wouldn't be marked on the map, and I wished that Shirin was there to help me. As I was struggling to find the place on the map, I remembered Shirin, long ago, marking a map to show where London was, and tears came to my eyes. There were some giggles in the class and I realized that some of the boys in my class were laughing at me. I ran to my desk and put my head down, I didn't want to show my face to anyone. I heard their chants: "Cry-baby! Cry-baby!" Miss Richards tried to quieten them down, but the noise grew and grew. I could also hear Maxine's voice: "Please leave Nazia alone! Just shut up, you lot!"

Luckily, the bell for end of school went; for a moment the noise dropped slightly. Then Miss Richards called out: "Make sure you put your chairs up before you leave!" The noise increased: there was a clattering of chairs on desks, whistles and shouts, and a rush towards the door.

Miss Richards turned to me: "They are immature, don't take any notice of them. I'll have to talk to them in form-period." Then she paused and helped me to wipe my eyes and face: "I am sure you will hear from Shirin soon, I'm sure she's all right." I dried my eyes and went back to my desk to collect my bag.

Rukeya and Maxine had waited for me outside. As I came out, they joined me and the three of us walked, arm in arm, across the playground.

Sister

BY

CAROL

SHIELDS

Curious
the way our mother's
gestures survive
in us.

When she was alive
we never noticed
but now in the dark
opening up
since her sudden
leaving, we are more aware.

A thousand miles away in
a similar kitchen
you pause
to lift a coffee cup.

And here
my smaller identical wrist
traces the same arc,
precise in mid-morning air,

linking us together,
reminding us
exactly who she was,
who we are.

The Secret World of Siblings

BY

ERICA E. GOODE

hey have not been together like this for years, the three of them standing on the close-cropped grass. New England lawns and steeples spread out below the golf course. He is glad to see his older brothers, has always been glad to have "someone to look up to, to do things with." Yet he also knows the silences between them, the places he dares not step, even though they are all grown men now. They move across the greens, trading small talk, joking, But at the 13th hole, he swings at the ball, duffs it and his brothers begin to needle him. "I should be better than this," he thinks. Impatiently, he swings again, misses, then angrily grabs the club and breaks it in half across his knee. Recalling this outburst later, he explains, simply, "They were beating me again."

As an old man, Leo Tolstoy once opined that the simplest relationships in life are between brother and sister. He must have been delirious at the time. Even lesser mortals, lacking Tolstoy's acute eye and literary skill, recognize the power of the word *sibling* to reduce normally competent, rational human beings to raw bundles of anger, love, hurt, longing and disappointment—often in a matter of minutes. Perhaps they have heard two elderly sisters dig at each other's sore spots with astounding accuracy, much as they did in junior high. Or have seen a woman corner her older brother at a family reunion, finally venting 30 years of pent-up resentment. Or watched remorse and yearning play across a man's face as he speaks of the older brother whose friendship was chased away long ago, amid dinner table taunts of "Porky Pig, Porky Pig, oink, oink, oink!"

Sibling relationships—and 80 percent of Americans have at least one—outlast marriages, survive the death of parents, resurface after quarrels that would sink any friendship. They flourish in a thousand incarnations of closeness and distance, warmth, loyalty and distrust. Asked to describe them, more than a few people stammer and hesitate, tripped up by memory and sudden bursts of unexpected emotion.

Traditionally, experts have viewed siblings as "very minor actors on the stage of human development," says Stephen Bank, Wesleyan University psychologist and co-author of *The Sibling Bond.* But a rapidly expanding body of research is showing what goes on in the playroom or in the kitchen while dinner is being cooked exerts a profound influence on how children grow, a contribution that approaches, if it may not quite equal, that of parenting. Sibling relationships shape how people feel about themselves, how they understand and feel about others, even how much they achieve. And more often than not, such ties represent the lingering thumbprint of childhood upon adult life, affecting the way people interact with those closest to them, with friends and co-workers, neighbors and spouses—a topic explored by an increasing number of popular books, including *Mom Loved You Best,* the most recent offering by Dr. William and Mada Hapworth and Joan Heilman.

Shifting Landscape

In a 1990s world of shifting social realities, of working couples, disintegrating marriages, "blended" households, disappearing grandparents and families spread across a continent, this belated validation of the importance of sibling influences probably comes none too soon. More and more children are stepping in to change diapers, cook meals and help with younger siblings' homework in the hours when parents are still at the office. Baby boomers, edging into middle age, find themselves squaring off once again with brothers and sisters over the care of dying parents or the division of inheritance. And in a generation where late marriages and fewer children are the norm, old age may become for many a time when siblings—not devoted sons and daughters—sit by the bedside.

It is something that happened so long ago, so silly and unimportant now that she is 26 and a researcher at a large, downtown office and her younger brother is her best friend, really, so close that she talks to him at least once a week. Yet as she begins to speak, she is suddenly a 5-year-old again on Christmas morning, running into the living room in her red flannel pajamas, her straight blond hair in a ponytail. He hasn't even wrapped it, the little, yellow-flowered plastic purse. Racing to the tree, he brings it to her, thrusts it at her—"Here's your present, Jenny!"—smiling that stupid, adoring, little brother smile. She takes the purse and hurls it across the room. "I don't want your stupid present," she yells. A small crime, long ago forgiven. Yet she says, "I still feel tremendously guilty about it."

Sigmund Freud, perhaps guided by his own childhood feelings of rivalry, conceived of siblingship as a story of unremitting jealousy and competition. Yet, observational studies of young children, many of them the groundbreaking work of Pennsylvania State University psychologist Judy Dunn and her colleagues, suggest that while rivalry between brothers and sisters is common, to see only hostility in sibling relations is to miss the main show. The arrival of a younger sibling may cause distress to an older child accustomed to parents' exclusive attention, but it also stirs enormous interest, presenting both children with the opportunity to learn crucial social and cognitive skills: how to comfort and empathize with another person, how to make jokes, resolve arguments, even how to irritate. The lessons in this life tutorial take as many forms as there are children and parents....

Parental Signals

To some extent, parents set the emotional tone of early sibling interactions. Dunn's work indicates, for example, that children whose mothers encourage them to view a newborn brother or sister as a human being, with needs, wants and feelings, are friendlier to the new arrival over the next year, an affection that is later reciprocated by the younger child. The quality of parents' established relationships with older siblings can also influence how a new younger brother or sister is received. In another of Dunn's studies, first-born daughters who enjoyed a playful, intense relationship with their mothers treated newborn siblings with more hostility, and a year later the younger children were more hostile in return. In contrast, older daughters with more contentious relationships with their mothers greeted the newcomer enthusiastically—perhaps relieved to have an ally. Fourteen months later, these older sisters were more likely to imitate and play with their younger siblings and less apt to hit them or steal their toys.

In troubled homes, where a parent is seriously ill, depressed or emotionally unavailable, siblings often grow closer than they might in a happier environment, offering each other solace and protection. This is not always the case, however. When parents are on the brink of separation or have already divorced and remarried, says University of Virginia psychologist E. Mavis Hetherington, rivalry between brothers and sisters frequently increases, as they struggle to hold on to their parents' affection in the face of the breakup. If anything, it is sisters who are likely to draw together in a divorcing family, while brothers resist forming tighter bonds. Says Hetherington: "Males tend to go it alone and not to use support very well."

Much of what transpires between brothers and sisters, of course, takes place when parents are not around. "Very often the parent doesn't see the subtlety or the full cycle of siblings' interactions," says University of Hartford psychologist Michael Kahn. Left to their own devices, children tease, wrestle, and play make-believe. They are the ones eager to help pilot the pirate ship or play storekeeper to their sibling's impatient customer. And none of this pretend play, researchers find, is wasted. Toddlers who engage regularly in make-believe with older siblings later show a precocious grasp of others' behavior. Says Dunn: "They turn out to be the real stars at understanding people."

Obviously, some degree of rivalry and squabbling between siblings is natural. Yet in extreme cases, verbal or physical abuse at the hands of an older brother or sister can leave scars that last well into adulthood. Experts like Wesleyan University's Bank distinguish between hostility that takes the form of humiliation or betrayal and more benign forms of conflict. From the child's perspective, the impact of even normal sibling antagonism may depend in part on who's coming out ahead. In one study, for example, children showed higher self-esteem when they "delivered" more teasing, insults and other negative behaviors to their siblings than they received. Nor is even intense rivalry necessarily destructive. Says University of Texas psychologist Duane Buhrmester: "You may not be happy about a brother or sister who is kind of pushing you along, but you may also get somewhere in life."

They are two sides of an equation written 30 years ago: Michèle, with her raven-black hair, precisely made-up lips, restrained smile: Arin, two years older, her easy laugh filling the restaurant, the sleeves of her gray turtleneck pulled over her hands.

This is what Arin thinks about Michèle: "I have always resented her, and she has always looked up to me. When we were younger, she used to copy me, which would drive me crazy. We have nothing in common except our family history—isn't that terrible? I like her spirit of generosity, her direction and ambition. I dislike her vapid conversation and her idiotic friends. But the reality is that we are very close, and we always will be."

This is what Michèle sees: "Arin was my ideal. I wanted to be like her, to look like her. I think I drove her crazy. Once, I gave her a necklace I thought was very beautiful. I never saw her wear it. I think it wasn't good enough, precious enough. We are so different—I wish that we could be more like friends. But as we get older, we accept each other more."

It is something every brother or sister eventually marvels at, a conundrum that novelists have played upon in a thousand different ways: There are two children. They grow up in the same house, share the same parents, experience many of the same events. Yet they are stubbornly, astonishingly different.

A growing number of studies in the relatively new field of behavioral genetics are finding confirmation for this popular observation. Children raised in the same family, such studies find, are only very slightly more similar to each other on a variety of personality dimensions than they are, say, to Bill Clinton or to the neighbor's son. In cognitive abilities, too, siblings appear more different than alike. And the extent to which siblings *do* resemble one another in these traits is largely the result of the genes they share—a conclusion drawn from twin studies, comparisons of biological siblings raised apart and biological children and adopted siblings raised together.

Contrasts

Heredity also contributes to the *differences* between siblings.

About 30 percent of the dissimilarity between brothers and sisters on many personality dimensions can be accounted for by differing genetic endowments from parents. But that still leaves 70 percent that cannot be attributed to genetic causes, and it is this unexplained portion of contrasting traits that scientists find so intriguing. If two children who grow up in the same family are vastly different, and genetics accounts for only a minor part of these differences, what else is going on?

The answer may be that brothers and sisters don't really share the same family at all. Rather, each child grows up in a unique family, one shaped by the way he perceives other people and events, by the chance happenings he alone experiences, and by how other people—parents, siblings and teachers—perceive and act toward him. And while for decades experts in child development have focused on the things that children in the same family share—social class, child-rearing attitudes and parents' marital satisfaction, for example—what really seem to matter are those things that are not shared. As Judy Dunn and Pennsylvania State behavioral geneticist Robert Plomin write in *Separate Lives: Why Siblings Are So Different,* "Environmental factors important to development are those that two children in the same family experience differently."

Asked to account for children's disparate experiences, most people invoke the age-old logic of birth order. "I'm the middle child, so I'm cooler headed," they will say, or "Firstborns are high achievers." Scientists, too, beginning with Sir Francis Galton in the 19th century, have sought in birth order a way to characterize how children diverge in personality, IQ or life success. But in recent years, many researchers have backed away from this notion, asserting that when family size, number of siblings and social class are taken into account, the explanatory power of birth ranking becomes negligible. Says one psychologist: "You wouldn't want to make a decision about your child based on it."

At least one researcher, however, argues that birth order does exert a strong influence on development, particularly on attitudes toward authority. Massachusetts Institute of Technology historian Frank Sulloway, who has just completed a 20-year analysis of 4000 scientists from Copernicus through the 20th century, finds that those with older siblings were significantly more likely to have contributed to or supported radical scientific revolutions, such as Darwin's theory of evolution. Firstborn scientists, in contrast, were more apt to champion conservative scientific ideas. "Later-borns are consistently more open-minded, more intellectually flexible and therefore more radical," says Sulloway, adding that later-borns also tend to be more agreeable and less competitive.

Hearthside Inequities

Perhaps most compelling for scientists who study sibling relationships are the ways in which parents treat their children differently and the inequalities children perceive in their parents' behavior. Research suggests that disparate treatment by parents can have a lasting effect, even into adulthood. Children who receive more affection from fathers than their siblings do, for example, appear to aim their sights higher in terms of education and professional goals, according to a study by University of Southern California psychologist Laura Baker. Seven-year-olds treated by their mothers in a less affectionate, more controlling way than their brothers or sisters are apt to be more anxious and depressed. And adolescents who say their parents favor a sibling over themselves are more likely to report angry and depressed feelings.

Parental favoritism spills into sibling relationships, too, sometimes breeding the hostility made famous by the Smothers Brothers in their classic 1960s routine, "Mom always loved you best." In families where parents are more punitive and restrictive toward one child, for instance, that child is more likely to act in an aggressive, rivalrous and unfriendly manner toward a brother or sister, according to work by Hetherington. Surprisingly, it may not matter who is favored. Children in one study were more antagonistic toward siblings even when *they* were the ones receiving preferential treatment.

Many parents, of course, go to great lengths to distribute their love and attention equally. Yet even the most consciously egalitarian parenting may be seen as unequal by children of different ages.... Adolescents report favoritism when their mothers and fathers insist that none exists. Some parents express surprise that their children feel unequally treated, while at the same time they describe how one child is more demanding, another needs more discipline. And siblings almost never agree on their assessments of who, exactly, Mom loves best.

Nature vs. Nurture

Further complicating the equation is the contribution of heredity to temperament, each child presenting a different challenge from the moment of birth. Plomin, part of a research team led by George Washington University psychiatrist David Reiss that is studying sibling pairs in 700 families nationwide, views the differences between siblings as emerging from a complex interaction of nature and nurture. In this scheme, a more aggressive and active child, for example, might engage in more conflict with parents and later become a problem child at school. A quieter, more timid child might receive gentler parenting and later be deemed an easy student.

In China, long ago, it was just the two of them, making dolls out of straw together in the internment camp, putting on their Sunday clothes to go to church with their mother. She mostly ignored her younger sister, or goaded her relentlessly for being so quiet. By the time they were separated—her sister sailing alone at 13 for the United States—there was already a wall between them, a prelude to the stiff Christmas cards they exchange, the rebuffed phone calls, the impersonal gifts that arrive in the mail.

Now, when the phone rings, she is wishing hard for a guardian angel, for someone to take away the pain that throbs beneath the surgical bandage on her chest, keeping her curled under the blue and white cotton coverlet. She picks up the receiver, recognizes her sister's voice instantly, is surprised, grateful, cautious all at once. How could it be otherwise after so many years? It is the longest they have spoken in 50 years. And across the telephone wire, something is shifting, melting in the small talk about children, the wishes for speedy recovery. "I think we both realized that life can be very short," she says. Her pain, too, is dulling now, moving away as she listens to her sister's voice. She begins to say a small prayer of thanks....

Adult Bonds

Rivalry between siblings wanes after adolescence, or at least adults are less apt to admit competitive feelings. Strong friendships also become less intense, diluted by geography, by marriage, by the concerns of raising children and pursuing independent careers. In national polls, 65 percent of Americans say they would like to see their siblings more than the typical "two or three times a year."... Yet for some people, the detachment of adulthood brings relief, an escape from bonds that are largely unwanted but never entirely go away. Says one woman about her brothers and sisters: "Our values are different, our politics diametrically opposed. I don't feel very connected, but there's still a pressure to keep up the tie, a kind of guilt that I don't have a deeper sense of kinship."...

Warmth or Tolerance

Given the mixed emotions many adults express about sibling ties, it is striking that in national surveys the vast majority—more than 80 percent—deem their relationships with siblings to be "warm and affectionate." Yet this statistic may simply reflect the fact that ambivalence is tolerated more easily at a distance, warmth and affection less difficult to muster for a few days a year than on a daily basis. Nor are drastic breaches between siblings—months or years of silence, with no attempt at rapprochement—unheard of....

Sibling feuds often echo much earlier squabbles and are sparked by similar collisions over shared responsibility or resources—who is doing more for an ailing parent, how inheritance should be divided. Few are long lasting, and those that are probably reflect more severe emotional disturbance. Yet harmonious or antagonistic patterns established in childhood make themselves felt in many adult's lives. Says psychologist Kahn: "This is not just kid stuff that people outgrow." One woman, for example, competes bitterly with a slightly older co-worker, just as she did with an older brother growing up. Another suspects that her sister married a particular man in part to impress her. A scientist realizes that he argues with his wife in exactly the same way he used to spar with an older brother.

For most people, a time comes when it makes sense to rework and reshape such "frozen images" of childhood—to borrow psychologist Bank's term—into designs more accommodating to adult reality, letting go of ancient injuries, repairing damaged fences. In a world of increasingly tenuous family connections, such renegotiation may be well worth the effort. Says author Judith Viorst, who has written of sibling ties: "There is no one else on Earth with whom you share so much personal history."

Brothers

～

BY

BRET

LOTT

This much is fact:

There is a home movie of the two of us, sitting on the edge of the swimming pool at my grandma and grandpa's old apartment building in Culver City. The movie, taken some time in early 1960, is in color, though the color has faded, leaving my brother Brad and me milk-white and harmless children, me a year and a half old, Brad almost four. Our mother, impossibly young, is in the movie, too. She sits next to me, on the right of the screen. Her hair, for all the fading of the film, is coal black, shoulder length and parted in the middle, curled up on the sides. She has on a bathing suit covered in purple and blue flowers, the color in them nearly gone. Next to me, on the left of the screen, is Brad in his white swimming trunks, our brown hair faded to only the thought of brown hair. I am in the center, my fat arms up, bent at the elbows, fingers curled into fists, my legs kicking away at the water, splashing and splashing. I am smiling, the baby of the family, the center of the world at that very instant, though my mother is pregnant, my little brother Tim some six or seven months off, my little sister Leslie, the last child, still three years distant. The pool water before us is only a thin sky blue, the bushes behind us a dull and lifeless light green. There is no sound.

My mother speaks to me, points at the water, then looks up. She lifts a hand to block the sun, says something to the camera. Her skin is the same white as ours, but her lips are red, a sharp cut of lipstick moving as she speaks.

I am still kicking. Brad is looking to his right, off the screen, his feet in the water, too, but moving slowly. His hands are on the edge of the pool, and he leans forward a little, looks down into the water.

My mother still speaks to the camera, and I give an extra hard kick.

~

Brad flinches at the water, squints his eyes, while my mother laughs, puts a hand to her face. She looks back to the camera, keeps talking, a hand low to the water to keep more from hitting her. I still kick hard, still send up bits of water, and I am laughing a baby's laugh, mouth open and eyes nearly closed, arms still up, fingers still curled into fists.

More water splashes at Brad, who leans over to me, says something. Nothing about me changes: I only kick, laugh.

He says something again, his face leans a little closer to mine. Still I kick.

This is when he lifts his left hand from the edge of the pool, places it on my right thigh, and pinches hard. It's not a simple pinch, not two fingers on a fraction of skin, but his whole hand, all his fingers grabbing the flesh just above my knee, and squeezing down hard. He grimaces, his eyes on his hand, on my leg.

And this is when my expression changes, of course: in an instant I go from a laughing baby to a shocked one, my mouth a perfect O, my body shivering so that my legs kick even harder, even quicker, but just this one last time. They stop, and I cry, my mouth open even more, my eyes all the way closed. My hands are still in fists.

Then Brad's hand is away, and my mother turns from speaking to the camera to me. She leans in close, asking, I am certain, what's wrong.

The movie cuts then to my grandma, white skin and silver hair, seated on a patio chair by the pool, above her a green and white striped umbrella. She has a cigarette in one hand, waves off the camera with the other. Though she died eight years ago, and though she, too, loses color with each viewing, she is still alive up there, still waves, annoyed, at my grandpa and his camera, the moment my brother pinched hell out of me already gone.

~

This much is fact, too:

Thumbtacked to the wall of my office is a photograph of Brad and me, taken by my wife in November 1980, the date printed on the border. In it we stand together, I a good six inches taller than he, my arm around his shoulder. The photograph is black and white, as though the home movie and its sinking colors were a prophecy, pointed to this day twenty years later: we are at the tidepools at Portuguese Bend, out on the Palos Verdes Peninsula: in the background are the stone-gray bluffs, to the left of us the beginnings of the black rocks of the pools, above us the perfect white of an overcast sky.

Brad has on a white Panama hat, a gray hooded sweatshirt, beneath it a collarless shirt. His face is smooth-shaven, and he is grinning, lips together, eyes squinted nearly shut beneath the brim of the hat. It is a goofy smile, but a real one.

I have on a cardigan with an alpine design around the shoulders, the rest of it white, the shawl collar on it black, though I know it to have been navy blue. I have on a button-down Oxford shirt, sideburns almost to my earlobes. I have a mustache, a pair of glasses too large for my face, and I am smiling, my mouth open to reveal my big teeth. It isn't my goofy smile, but a real one, too.

~

These are the facts of my brother: the four-year-old pinching me, the twenty-four-year-old leaning into me, grinning.

But between the facts of these two images lie twenty years of the play of memory, the dark and bright pictures my mind has retained, embroidered upon, made into things they are, and things they are not. There are twenty years of things that happened between my brother and me, from the fist-fight we had in high school over who got the honey-bun for breakfast, to his phone call to me from a tattoo parlor in Hong Kong, where he'd just gotten a Chinese junk stitched beneath the skin of his right shoulder blade; from his showing me one summer day how to do a death drop from the jungle gym at Elizabeth Dickerson Elementary, to his watching while his best friend and our next-door neighbor, Lynn Tinton, beat me up on the driveway of our home, a fight over whether I'd fouled Lynn at basketball. I remember—memory, no true picture, certainly, but only what I have made the truth by holding tight to it, playing it back in my head at will and in the direction I wish it to go—I remember lying on my back, Lynn's knees pinning my shoulders to the driveway while he hit my chest, and looking up at Brad, the basketball there at his hip, him watching.

I have two children now. Both boys, born two and a half years apart.

I showed the older one, Zeb—he is almost eight—the photograph, asked him who those two people were.

He held it in his hands a long while. We were in the kitchen. The bus comes at seven-twenty each morning, and I have to have lunches made, breakfasts set out, all before that bus comes, and before Melanie takes off for work, Jacob in tow, to be dropped off at the Montessori school on her way to her office.

I waited, and waited, finally turned from him to get going on his lunch.

"It's you," he said. "You have a lot of hair," he said.

"Who's the other guy?" I said.

I looked at him, saw the concentration on his face, the way he brought the photograph close, my son's eyes taking in his uncle as best he could.

He said, "I don't know."

"That's your Uncle Brad," I said. "Your mom took that picture ten years ago, long before you were ever born."

He still looked at the picture. He said, "He has a beard now."

I turned from him, finished with the peanut butter, now spread jelly on the other piece of bread. This is the only kind of sandwich he will eat at school.

He said from behind me, "Only three years before I was born. That's not a long time."

I stopped, turned to him. He touched the picture with a finger.

He said, "Three years isn't a long time, Dad."

But I was thinking of my question: *Who's the other guy?* and of the truth of his answer: *I don't know.*

Zeb and Jake fight.

They are only seven and a half and five, and already Zeb has kicked out one of Jake's bottom teeth. Melanie and I were upstairs wrapping Christmas presents in my office, a room kept locked the entire month of December because of the gifts piled up in there.

We heard Jake's wailing, dropped the bucket of Legos and the red and green Ho! Ho! Ho! paper, ran for the hall and down the stairs.

There in the kitchen stood my two sons, Jacob with his eyes wet, whimpering now, a hand to his bottom lip.

I made it first, yelled, "What happened?"

"I didn't do it," Zeb said, backing away from me there with my hand to Jacob's jaw. Melanie stroked Jacob's hair, whispered, "What's wrong?"

Jacob opened his mouth then, showed us the thick wash of blood between his bottom lip and his tongue, a single tooth, horribly white, swimming up from it.

"We were playing Karate Kid," Zeb said, and now he was crying. "I didn't do it," he said, and backed away even farther.

One late afternoon a month or so ago, Melanie came home with the groceries, backed the van into the driveway to make it easier to unload all those plastic bags. When we'd finished, we let the boys play outside, glad for them to be out of the kitchen while we sorted through the bags heaped on the counter, put everything away.

Melanie's last words to the two of them, as she leaned out the front door into the near-dark: "Don't play in the van!"

Not ten minutes later Jacob came into the house, slammed shut the front door like he always does. He walked into the kitchen, his hands behind him. He said, "Zeb's locked in the van." His face takes on the cast of the guilty when he knows he's done something wrong: his mouth was pursed, his eyebrows up, his eyes looking right into mine. He doesn't know enough yet to look away. "He told me to come get you."

He turned, headed for the door, and I followed him out onto the porch where, before I could even see the van in the dark, I heard Zeb screaming.

I went to the van, tried one of the doors. It was locked, and Zeb was still screaming.

"Get the keys!" he was saying. "Get the keys!"

I pressed my face to the glass of the back window, saw Zeb inside jumping up and down. "My hand's caught," he cried.

I ran into the house, got the keys from the hook beneath the cupboard, only enough time for me to say to Melanie, "Zeb's hand's closed in the back door," and turned, ran back out.

I made it to the van, unlocked the big back door, pushed it up as quick as I could, Melanie already beside me.

Zeb stood holding the hand that'd been closed in the door. Melanie and I both took his hand, gently examined the skin, wiggled fingers, and in the dull glow of the dome light we saw that nothing'd been broken, no skin torn. The black foam lining the door had cushioned his fingers, so that they'd only been smashed a little, but a little enough to scare him, and to make blue bruises there the next day.

But beneath the dome light there'd been the sound of his weeping, then the choked words, "Jacob pulled the door down on me."

From the darkness just past the line of light from inside the van came my second son's voice: "I didn't do it."

I have no memory of the pinch Brad gave me at the edge of an apartment complex pool, no memory of my mother's black hair—now it's sort of brown—nor even any memory of the pool itself. There is only that bit of film.

But I can remember putting my arm around his shoulder, leaning into him, the awkward and alien comfort of that touch. In the photograph we are both smiling, me a newlywed with a full head of hair, he only a month or so back from working a drilling platform in the Gulf of Mexico. He'd missed my wedding six months before, stranded on the rig, he'd told us, because of a storm.

What I believe is this: that pinch was entry into our childhood; my arm around him, our smiling, the proof of us two surfacing, alive but not unscathed.

And here are my own two boys, already embarked.

My Sister's Hand

~

BY

JILL S.

RINEHART

I hold tight to my sister's hand as I help her wind her way up the twisting climber to the top of the slide. Another step up.

"Careful, Ali," I say as she makes an unsteady lean to her left, away from me. She is seven years old, but people think she is about four. She is small and her once-blonde hair is as dark as mine now. Her blue eyes shine large behind her tortoiseshell glasses and her hearing aids are quietly tucked behind her ears, covered by her neatly cropped hair. I am fourteen years old—a giant amidst this rowdy group of Saturday afternoon children.

"What's her problem?" one little boy asks me. His Langford Park T-shirt is torn at the shoulder. Problem? I look at Alison's knees. Good, no blood. Her nose isn't running, her shoes are on, and her dress is still covering her underwear.

"Nothing," I say to the boy, helping Ali up the top step.

"Why is she walking like that?" He is genuinely curious, as if she could have some type of cold he might get. Why? I shuffle through the words in my mind: undiagnosed neurological movement disorder...possible brain stem damage...metabolic deficiency...moderate retardation… hearing loss.

"She's deaf," I say.

The boy nods in understanding. Deafness, I reason, is a trendy handicap. Linda of "Sesame Street" is deaf. She talks with her hands and Big Bird understands her.

Alison smiles back at me and hangs on the crossbar at the top of the slide, trying to kick her legs up like the kids had done before us. She gets one leg up and then the other, but not both at the same time. Then she stands there looking down at the boy's torn shirt. "Well, hurry up!" the boy yells to us, waving his arms like a traffic controller. I motion for her to sit down in my lap and tuck my red Nikes under her legs. We laugh as we fly down the slide.

Six weeks after my sister was born, she began to shake uncontrollably. Her tiny arms and legs twitched as if from cold or fear and her tongue clicked against the roof of her mouth. At first the doctors thought she had epilepsy. No problem; with a few doses of Tegretol a day she would be perfectly fine. Yet, even with her larynx trembling so that her cries would vibrate in an eerie wave of urgency, when hooked up to an EEG, her brain activity appeared to be absolutely normal.

I am home with Alison, our parents are out for the evening. I have to write a story for my seventh grade English class tomorrow. Every time I look at my paper I see the long, loud, locker-lined halls of my junior high. My teacher says we have to read the stories to the class. I am struck by a wave of panic.

Alison is sitting in front of me on the floor, watching TV. Her face is about six inches from the screen. A commercial comes on, interrupting "M*A*S*H*," and she looks back at me. I think she wants popcorn. She sees me crying on the couch and walks over to me, bumping her knees on the coffee table and just missing the burning wood stove by one unsteady pace. "Jill's crying," she says. She brushes her hand across my tears and she hugs me. "Oh it's OK, Jill. It's OK."

In fact, almost every neural diagnostic test Alison has been subjected to has been normal. Yet her learning abilities are obviously delayed and her physical stability fluctuates from day to day. Some days her gait is fairly steady, other days she needs someone to help her balance as she walks, and occasionally she is unable to walk at all. Stress from sickness or exhaustion tends to exacerbate her unsteadiness and makes her more likely to twitch. We also don't know how much it is possible for Alison to learn. Intelligence is too difficult to gauge. So we were pleased when, after seven years, we discovered that Alison had substantial hearing loss. It was the answer to almost everything. Her slow learning, her slurred speech, even why she called our dad "Jer" instead of "Daddy" (she had been able to hear Mom yell down to the garage, "Hey, Jer?" but not the softly spoken "Daddy"). We now had something tangible, something "treatable."

Alison sits beside me in my bed. We share a room because she is afraid to sleep alone. We are working on her colors and ABC's. She is nine and I am sixteen. On page one is a big red apple, with a big red letter "A."

"What's this?" I ask. She says, "Appo."

"Good! What's this?" I ask.

"A for Alison!!" she exclaims, smiling and signing her name by folding her hand into the letter sign "A" and circling it before her face in the sign motion for "beautiful."

We discovered that these "spells" (as we call them) stop with sleep. The twitching episodes interfere with her ability to fall asleep on her own, but a small dose of a sedative will help her relax. The twitching leaves her, little by little, with each breath closer to sleep.

I leave for college today. The cool hospital corridors are unpleasant despite the ninety-degree heat outside. Alison has been here this time for about ten days. Her right hand has been twitching for three weeks. The twitching rolls through her in waves of increasing intensity. It slows, teasingly, for one or two hours, but never completely stops. The hand is so swollen that light barely passes between her now bluish fingers. The doctors think she maybe broke it during a more violent spell, but her x-rays are normal.

She lies in her hospital bed, her eyes not focusing well and her fine, brown hair damp with sweat. I say goodbye to her. I am leaving her for the first time in her eleven years. She doesn't understand that I won't be back tomorrow. She doesn't know about college. My mother and I have started a photo book story called "Jill Goes to College" so that she can understand. Mother will explain everything when Alison gets better.

Mother says goodbye to me, her eyes on her other daughter as she goes into status—a critical seizure. Ali is no longer aware of herself, no longer just twitching. Her body contorts as misguided electrical impulses pass through her brain. The doctor decides to induce Alison into a coma, the deepest level of sleep, hoping this will be enough to stop the seizure and bring Alison back.

The spells almost always stop with sleep. But one dreadfully hot summer, when Alison was eleven, we tried to experiment with her medications, thinking we had discovered a drug that might help her. It did not. Instead, the new drug caused the levels of the medications that were helping control her movements to drop dramatically. These medications—her only defense—no longer protected her. The depth of her disorder, preciously masked by these medications, was revealed to us. Ali's hand, and indeed her whole left side, twitched visibly for weeks. A muscle biopsy was done on her right thigh (the non-twitching side) to have fibroblasts grown for some genetic tests. At the very first layer of muscle, Alison's right leg was twitching.

~

It is spring break my freshman year in college and I'm in Tennessee training for sprints, dreadfully homesick. Three months since I've been home, three more 'til I get home. During the two weeks of training, I get some thirty photos from home: my brother with his train, my mother's handmade quilt, our friend's new car, and my sister—a pale, thin Alison. Her hair has grown out some and a small ponytail is tied to one side of her head with a pink ribbon that matches her glasses. She sits in a gold wingback chair with her tights bunching around her knees. She is resting her chin on her hand and her bony, cocked wrist doesn't look strong enough to support her head. She strains to hold up her chin, but her eyes shine clearly. On the back of the photo opposite the stamp is written "Beautiful Alison."

Alison came home from the hospital at the end of September without biotin and with readjusted medication levels. While at the hospital she had developed pneumonia from an aspiration (an inhalation of mucus into the lung, which probably occurred during a spell) and they decided she was better off at home.

She wasn't able to go to school that fall and remained at home through most of the winter. A nurse came to the house a couple of times a week. He helped her regain lost muscle tone through physical therapy. A teacher from school also visited Ali twice a week. Alison went back to school in the spring, somewhat stronger but still often dependent on a wheelchair.

Christmastime, 1988: I am home for break and leave for France today. I tell Alison I am going back to school and ask if she wants to go to the airport with me. I reach for her hand. "No," she says and signs, "Sleep now." She turns away from me and heads into my parents' bedroom. I go to her and sit by her side on the bed. "G'away," she says. It is too much and I turn to leave. Mother steps by me and signs to Alison, "Ali don't sleep now, come to the airport and see the airplanes." Alison is crying and I hug her until we are all three in tears. She understands college.

Presently thirteen, Alison is stable. She lost a lot of weight while she was in the hospital, her speech seemed less articulate, and she slept about sixteen hours a day. However, she is now walking better, her signing skills are improving, and she continues to learn. She still has outbreaks of twitching. Just after Thanksgiving 1989, she started at a boarding school. She attends the Minnesota State Academy for the Deaf located in Faribault, about an hour from home. Here we found a facility that is willing to adapt to Alison's needs, even more so than her old school. She is one of the one hundred and sixty deaf students between kindergarten and twelfth grade that attend the academy.

I am home for the fall, driving down to Faribault, with two cans of Coke and a box of Doo Dads—Alison's favorites. I am anxious to see her at her new school.

I find Alison in the playroom working on a puzzle with her roommate. She gives me a blank look. "No," she says, "I don't want to go home now. Stay at school." Mom warned me about this.

"What did she say?" the houseparent asks me.

"She thinks it is time to go home now because I am here, and she doesn't want to leave," I explain. The houseparent gives me a suspicious look and I shrug it off. "No, we don't beat her," I think to myself. I know Alison likes it just fine at home, she loves every one of us, and is happy when she is there. But she loves her newly gained independence.

Alison is stubborn with me. I offer her my hand to help her into the shower but she says, "No, do it myself." Everything is "myself." In sign language this is a hitch-hiker thumb tapped twice on the chest. Ali taps hers three times.

I thought of when I was thirteen, impossibly shy and suffering from nervous stomachaches. Going on the annual four-day school trip to the Boundary Waters was always a traumatic event. I could never have lived away from home at her age. Her strength amazes me.

Alison is the eighth multi-handicapped student enrolled in the Academy, but has a more complicated medical history than the others. She spends most of her school day with her class of five learning-disabled thirteen-year-olds. Her main goal is to improve communication skills. She is mainstreamed for part of the day into a class of deaf fifth graders. She has two hours of physical therapy a week. But her "real" day begins, like any other teenager's, outside of class. After school Alison plays wheelchair basketball, visits museums in the Twin Cities, attends MSAD sporting events and even Girl Scout meetings. She is gaining friends and enjoys visiting the older girls on the second floor of her dorm. No wonder Ali doesn't want to leave school.

In the morning, I fight off the temptation to help Alison with her clothes. I let her dress herself; it takes twenty minutes. In classes, I sit with her, translating her speech for the teachers only if they ask me to. She must sign more and learn to communicate on her own. In art class I let her spread her own glue and get it all over her hands. She walks herself to the bathroom and trips on the rug in the hall. I hear her stumble and I rush to block her fall, but realize she is fine without me. I watch as Ali starts to lift herself from the floor. Kneeling, she bends one leg out in front of her, and leans on the wall with one hand for support. She smiles up at me, laughing at herself, and reaches for my hand.

The teachers, administrators and houseparents are all a bit wary of Alison. They read her case history— "neurological disorder," "needs wheelchair," "full body twitching," "medically fragile…"—and under-standably, they get worried. But as they learn about Alison, her unrecord-ed history—invisible yet inextricably woven into her story—becomes apparent. Neurological disorder: aware of her own abilities. Needs wheelchair: increasingly self-supportive. Twitching: capable of the grace of a three-dimensional language. Medically fragile: strengthened by a lifetime of having to do things the hard way, harder than most people have to. Alison.

Shelly Niro *Portrait #4: Family: Stylistic Beadwork Designs and Blue Lights*

To My Dear and Loving Husband

~

BY

ANNE

BRADSTREET

If ever two were one, then surely we.
If ever man were lov'd by wife, then thee;
If ever wife was happy in a man,
Compare with me ye women if you can.
I prize thy love more than whole Mines of gold,
Or all the riches that the East doth hold.
My love is such that Rivers cannot quench,
Nor ought but love from thee, give recompense.
Thy love is such I can no way repay,
The heavens reward thee manifold, I pray.
Then while we live, in love lets so persever,
That when we live no more, we may live ever.

Spaghetti

BY

NANCY

BOTKIN

When my father ate spaghetti
he took his knife and fork and cut it
into little pieces, the knife
scraped the bottom of the plate,
a shrill sound which made my Italian
mother wince. The heaping pile
was reduced to mush and my mother
would scold him, then she showed him how
to do it. We all watched as she placed
her fork at the edge of her plate
and delicately twirled it until she
had a small, manageable amount which
she placed quietly in her mouth.
My Irish father said the hell with it
and he scooped it in, and we kids scooped
it in, shoveled it in. The silverware
clicked and flashed, a clash of metal
against metal, metal against teeth, metal
against china, and when the meal was over
the knives and forks lay in our plates
like exhausted soldiers, like surgeons' tools,
orange-red, sharp, dangerous as any weapons.

Heirloom

BY

EAMON

GRENNAN

Among some small objects I've taken from my mother's house
is this heavy, hand-size, cut-glass saltcellar. Its facets
find her at the diningroom table
reaching for the salt or passing it to my father
at the far end, his back to the window.

The table's a timebomb, ticking in the play of light
on the white cloth, a current of silence that holds us all
in our blind clutching after straws of talk: Father hidden
behind the newspaper, Mother filling our plates with food,
my own tongue blunted between them—the way
they couldn't seem to meet each other's eyes.

He'd leave the table early for an armchair, *just
a glance at the evening paper*, and she'd sit on until—
all small talk exhausted—we kids would clear things away,
stack dirty dishes by the sink in the scullery,
and store the saltcellar in the press
where it would absorb small tears of air
till the next time we'd need
its necessary bitter addition, as now it stands

on our kitchen table over here, is carried
to the diningroom for meals, a solid object
filling my hand, a presence at the centre of our talk,
its cheap cut glass outlasting flesh and blood
as heirlooms do. I take its salt
to the tip of my tongue, testing its savour
and spilling by chance
a tiny white hieroglyph of grains
which I pinch in my mother's superstitious fingers
and quick-scatter over my left shoulder,
keeping at bay and safe the darker shades.

Fathers

BY

LAKE

SAGARIS

There are fathers who arrive every night
in the subway—mothers
run smiling to meet them

There are fathers who sleep
in beds of earth and errors
—small, fatal betrayals—
their children
miss them
search for them
in every smile glued to a photo
in every fight at school
in every flower
covering a grave

There are also fathers who arrived
one time, or two, or more
but they left without knowing
the person hidden
in that animal childhood

And then
there are fathers like yours
who arrive once a week
or twice a month
or more
or less

You run laughing to hug him
your smile flying toward him
arms, eyes, knees
shining toward him

And he looks for you
in a truck
kept between visits
in a ball
bouncing down hallways
along streets, into streams
where you both bathed together
then left
he to his home
you to your mother's

These are fathers
who are always just arriving
tears dampening their pockets
small shadow of a child's palm
engraved in the cradle of their fists

His, Hers & Theirs

BY

LESLIE

BLAKE-CÔTÉ

~

The summer of '93—specifically, July 10—will always be memorable for Lois Dellert. That was her wedding day. As with any bride-to-be, last-minute preparations were hectic and exhausting. But unlike most new brides, Dellert had additional worries...

Would husband-to-be Yves Talbot's 10-year-old son Martin disrupt the party with spitballs as he had threatened? Or would he even show up?

Sound like something out of the *Dennis The Menace* movie?

Well, it's not. It's a wedding involving one of North America's fast growing phenomena—the stepfamily.

Martin is no Dennis the Menace. Rather, he's a confused and hurt child trying to understand why his father suddenly needed this new wife when father and son had been exceptionally close for so long. Nor is Lois the wicked stepmother of the fairy tales, but a 35-year-old woman, married once already, who loves her husband and wants to make a home for him, Martin and 19-year-old Adam, a stepson from his ex-wife's former marriage. This blended or stepfamily is the new look of the '90s. A couple and any combination of his, hers and theirs.

Statistics Canada figures show that in 1990, at least one partner in 30 per cent of marriages had been married before. In 1991, it was estimated that one in seven children in Canada was part of a stepfamily.

Lois Dellert and Yves Talbot are living proof that such a family can work and work well. But it takes time, much patience and goodwill—and lots of love. Dellert and Talbot have only known each other for a couple of years and are novices in this stepfamily business.

Before their wedding, Talbot told Martin: "I love you, but I won't tolerate your behavior. You have a choice to either enjoy the wedding with us or to not go."

In the end, all went well.

Dellert, formerly a forester from Victoria, B.C., and Talbot, former chief of family medicine at Mount Sinai Hospital, met in Montreal in 1991. A year later, Dellert moved to Toronto, where the family now lives.

Probably the two most challenging issues facing them are the long and close relationship of Talbot and Martin, and Lois's appearance as a rival for Yves's affection, attention and time. It's normal for kids to resent the intrusion as they see it and act out in an attempt to keep things as they were before.

"It's been very difficult for me to find my place, to know what my role is and what Martin's is," Dellert says.

Talbot recalls Martin asking, "So, Lois, are you going to be my mother? So what will you be? And what will I call you?"

But, he continues, "it was only when Martin suggested that Lois get an apartment that I realized how much he and I had a fused relationship."

None of this is news to Lillian Messinger, a marital therapist, senior social worker with the Clarke Institute of Psychiatry and author of the book *Remarriage—*

A Family Affair. Messinger is well aware of the fact that 47 per cent of first and 44 per cent of second marriages fail. Consequently, one of the first things a stepfamily must do is deal with the myths that abound, such as the one about "instant family—if I love you, of course I'll love your kids."

Not necessarily so, says Messinger. "It takes two to three years before a new stepfamily feels like a real family." Another reality is that the blended family is very different and definitely more complicated than a first-time family. The stepfamily, for example, has adults, at least one of whom has been married before. One or both of these adults bring children of varying ages and backgrounds together. These kids wind up living part of the time in two different households with all the turmoil and adjustment that such arrangements can bring.

Then there are the new grandparents, aunts, uncles and cousins. And a host of different values, customs, traditions and cultures. Dellert recognizes mistakes she has made, like coming in with both barrels blazing as the "reformer"—no more junk food, only healthy foods.

"The new partner cannot come in and feel that they will reorganize," Messinger says. "A lot of compromise is involved."

Dellert and Talbot chose to renovate Talbot's home for the family to live in. While this was economical, it has been difficult for Martin. Whenever possible, Messinger advises neither party should move into the other's home. This way, no one feels like a boarder. As well, kids can help make some decisions with things like decorating.

Donna and Joe Kirisits of Brampton did exactly that when they married in 1984 after living together since 1980.

"When we moved into a townhouse with Donna's two kids (David, then 12, and Dean, 9) we made sure the kids were involved in choosing the house and picking rooms and wallpaper," Joe recalls. "We didn't have tons of money but we had lots of fun."

Joe refers to David and Dean as "my boys. I always told them that I was not their natural father, that they already had a father." But, he adds, "they are my sons. I raised them. I wiped their noses, picked them up when they got into little street fights and was always there to encourage them. I brought them up and part of me is in them."

Did he love them instantly?

"I cared about them right away and I was trying to reassure them that their lives would be okay. And then love grows."

Though the boys are now 22 and 24 and live away from home, Donna says, "To this day they don't leave the house without hugging and kissing us goodbye."

But if Joe ever received a best father of the year award from David and Dean, he would be named worst father by his two daughters from a former marriage. When his first marriage ended, he found himself in constant conflicts over the kids, broken dates for visitation and myriad other stressful situations. As a result, he says, "I decided one week that it was too stressful and I was bringing all of this unhappiness into my *new* home. So I decided to stop seeing my daughters, then 8 and 3, even though I loved them and still do.

"For everything, you pay a price. I paid the price for the happiness of my new family and myself. I felt that I could have no impact on their lives and never would.

"My eldest, of course, is resentful and feels I abandoned her. And the boys missed having sisters. But the boys understand what I did and why I did it."

Ties

BY

SHIRLEY A.

SERVISS

My stepson is learning
to tie his yellow laces
around the legs of kitchen chairs,
as I used to do, underfoot
on my mother's tiled kitchen floor.

He helps me make Rice Krispie squares,
the recipe written in my childish hand
on scribbler paper, yellowed now,
stirs the melted marshmallows
with a wooden spoon, pats the cake
into the buttered pan.

Putting him to bed, I fan the sheets
up and down, the cool air
kissing him goodnight the way
my mother used to
tuck me in when I was ill.

I'm mothering from memory,
playing by ear, learning the skills
I need to be a parent
as he crosses one lace over
the other, pulls it through.
We're mastering knots
before we move on to bows.

'How We Adopted Me'

BY

JANE

MARKS

~

The April 11 cover story [of *The New York Times Magazine*] ("China's Market in Orphan Girls: How Li Sha Abandoned in Wuhan, Became Hanna Porter, Embraced in Greenwich Village") was great—right up until the last part. There, the author assumes that cross-racially adopted children don't ask questions until they're 6 or 7 and that explaining abandonment is simple as long as your child knows she's loved by the time she asks and you make it sound affirmative, as in "Your parents gave you up because they loved you." My husband, Bob, and I adopted Joshua Grant Marks when he was an infant named Dong-Young, abandoned in Seoul, Korea. It's been a joy to raise him, but the whole issue of abandonment has never been simple in our family and I'm not sure that it ever will be.

Our first sight of Josh with his chubby cheeks and black wispy hair like the top of a pineapple was in the photo our local adoption agency received from an orphanage in Seoul. Nothing was known about this baby but his name and birth date, which were written on a paper slip inside his clothes when he was found outside the Bukboo Police Station. Bob and I were overjoyed. We had a son!

When the red tape cleared and Josh was placed in my arms at Kennedy Airport, he was 3 months old and an adorable baby: vigorous, healthy and smart. He got used to us quickly and was so responsive that he seemed to understand and even share our pleasure in being his parents. As he grew, he could see that he didn't look like us. "I'm looking in the mirror and I'm seeing the face of a Korean boy," he observed solemnly when he was not quite 4. Even earlier, he pointed to an Asian man and said, "Mom, he's adopted, too." But there was nothing negative in Josh's mind. In fact, his favorite bedtime story was "How We Adopted Me." He knew it by heart and would begin, "We wanted me so badly...."

The part we had trouble telling him—that he had had another mother and father who gave him up—came out when Josh was 4 and I was pregnant with his brother, Chris. "Yes, you grew in a womb, but not mine," I admitted. "No!" Josh denied it. Gently, Bob and I explained to Josh that his birth parents loved him but couldn't take care of a baby, that we loved him and would never leave. Josh wasn't impressed. "I guess I must have been a bad baby," he concluded sadly. We explained that wasn't true.

What followed for Josh was a period of high anxiety. Josh refused to get out of the car at a children's fair. "A clown might steal me," he said, hiding his face and scrunching down in the seat. At home, he was unwilling to sleep with his window open. He thought that maybe his mother hadn't meant to give him up—a theory that made it hurt less—but it meant that a stranger was out there look-ing for him. We assured Josh that nobody would take him away and that his birth parents would be happy to know that he was loved and cared for.

But did he believe that? I spoke with Doris Mennone, the wonder-ful social worker who had done our adoptive home study. Was Josh's distress unique, I wondered. "Not in the least," she said firmly. "The questions in his mind—'Can I really stay here? Is this really where I belong? Is someone going to come and reclaim me?'—occur to most adopted children when they try to make sense of their situation. Kids really wonder. I know it's painful to hear it, but getting the pain out in the open helps to keep it from festering."

Gradually Josh's kidnapping fear went away, but not his wonder-ing. "Maybe *they* were my parents," he said months after that when an elderly Japanese couple stopped us in the street to ask directions. I realized then that Josh's early abandonment might never go away, no matter how many times we took him on our laps and assured him that his birth parents did love him and want him to have what they couldn't give him.

Not that Josh spent all his time brooding. He was a busy and gregarious little boy with a hundred different interests, from karate to Cub Scouts, horseback riding, endless games of G.I. Joe in our yard and even learning to play bagpipes. But underneath his outwardly happy, ordinary life, there was a certain poignancy. When Josh balked at passing outgrown toys and books along to Chris ("No, I want to save them for my children"), was it big-brother orneriness or an orphan's yearning to establish roots? And when he came home from school with pockets crammed full of abandoned metal doodads and even shards of concrete, was he being a pack rat or expressing something deeper as he explained with dignity, "What some people throw away as junk may be a treasure to someone else."

Josh had no trouble embracing our entire extended family as his own. He adored his grandparents in Minnesota and was proud of his middle name, Grant, knowing that his dad was related to a United States President. Josh also enjoyed "Fiddler on the Roof," understanding that the shtetl, Anatevka, reflected my Eastern European Jewish heritage. "*They* had to find a new home, too," he said when we saw the movie. Still, none of that helped when our pediatrician asked about Josh's family medical history or when a neighbor wondered out loud if Josh had any brothers or sisters in Korea.

"I hate not knowing anything," Josh complained. We tried to offer as a substitute what little we could surmise about those unknown parents: that they must have been handsome and terrific in so many ways because he was. Always, I emphasized their caring: the fact that Josh was left right outside the police station where he would be found immediately—and he was! (It was February, but newborn Josh didn't even catch a cold.) That had to count for a lot in a country where, we understood, it was against the law to abandon a baby. Josh needed to know that he could love both his mothers, that his story had no bad guys and, therefore, no victims, either.

Did I go overboard to reassure him? Sometimes, for sure! Though he loved his nursery school, I *had to be* the first mom there for pickup. (I couldn't bear for Josh to have to wonder, even for a moment, where *his* mother was.) When Loolabah, one of our cats, started urinating on everything, we still kept her—not because we loved her so much but because I was waiting, unwisely, for Josh's permission to give her away. I was afraid that he'd identify with her. Was that stupid? Of course he knew the difference between cats and people—or did he? In the end, it was Loolabah who solved the problem by peeing on Josh's backpack with his wallet inside. "She can go," Josh agreed. Still, there were tears when it came to the actual goodbye.

When Josh was 11 and Chris was 7, we sent both boys to their grandparents for two weeks, while we went to Europe. Josh was deeply homesick, and one day my father-in-law found him crying in the attic. "Grandpa," Josh sobbed, "how will I ever go to college?" Long after the boys returned home, Josh was bitter, accusing us of having "dumped" him even though he had fun and even though that separation (and surviving it) enabled him to go away to camp the next summer. To our relief, he loved the Spartan wilderness camp his cousin Ben had gone to at the same age. We could hardly wait to see Josh on visiting day, but after the briefest hug he skulked away. What was up? Kicking gravel, Josh explained, "I don't want to have to start missing you all over again."

Over the next few years, there were few if any visible signs that Josh was still struggling. He had metamorphosed from a daydreamer in class to an excellent student. He played sax in the band, acted in plays, sang in the chorus and was co-captain of his high-school fencing team. Rounding it out, he had a rich variety of wonderful friends, a lovely girlfriend, Karen, and even an assortment of jobs baby sitting and gardening. One night, Bob showed him an article in *Science Times,* exploring why Asian children excel at school. "It doesn't apply to me," Josh concluded, handing the paper back to his dad. "This is about *real* Asian kids with Asian parents who push them to work."

"Well," I said, "it might be hereditary. I'll bet your birth parents were smart."

"If they were, then why did they get rid of me?" Josh shot back. I went through the litany—that they were pressed and very poor. Perhaps his mother was very ill. Or maybe there wasn't a dad. Josh smirked. "Are you saying I'm a bastard?" he challenged.

"If you were born out of wedlock, it's no reflection on you—or them," I replied evenly. No further questions.

Suddenly, it seemed, he was a senior in high school and it was time to plan the trip that we had always wanted to take as a family to see the orphanage in Seoul. As Bob made the itinerary, we wondered if this trip would provide some closure for Josh or just reopen old wounds? I called our long-ago social worker to ask how it worked out for other adoptive families. "I don't have a clue," she admitted. "Everyone talks about it, but I don't know a family who's actually gone."

Well, we did go, and the experience was more rewarding than I'd dared to hope. Armed with the address of the orphanage (533-3 Sangmoon-dong, Tobong-ku) written out in Korean, we ventured forth. And what luck! Mr. Kim, who had founded that orphanage, the Korea Social Service receiving home, 28 years ago, was there and he generously spent the morning with us. Mr. Kim showed us the babies in the nursery and then in his office he called for Josh's file, which contained some baby pictures, different from the one we had seen so long ago. Mr. Kim offered one to us—a precious souvenir.

But that wasn't all. Mr. Kim spoke at length to Josh, emphasizing: "Your birth parents gave you a gift, and you should be grateful for the family life you have and a bright future." He went on to say that even though Korea is more affluent now than it was 19 years ago, the majority of Koreans still have a struggle. "Most of us are thin and tired," he said. "No time to enjoy life." He added that it's even tougher for orphans, who are cruelly stigmatized in Korea and are not even allowed to join certain army units. Josh was riveted. "I know that I'm lucky," he said with feeling. Mr. Kim hadn't told him much that we hadn't said for years. But clearly, it meant everything to hear that same story in a different voice.

We told Mr. Kim that we wanted to visit the Bukboo Police Station, too, where Josh was found. Mr. Kim shook his head and told us not to bother, adding, "I was Josh's first legal guardian and I say his life began right here." Josh looked thoughtful. "Yes," he agreed. "My life began right here." It was a special moment. In a sense, our son had let go of whatever and whomever might have come before.

Over the next week and a half, we drove around the country. Mr. Kim had told Josh to be proud of his Korean heritage and Josh agreed enthusiastically. "It's my homeland," he'd said. But away from cosmopolitan Seoul, we found that our Asian-born son with his American upbringing was a hybrid. He looked Korean, but he couldn't speak it, which confused people and frustrated Josh. "I feel like such a dork," he admitted, embarrassed after summoning a waitress in phrase-book Korean. (Had the waitress giggled?) "I guess I'm a foreigner," Josh said sadly. But he made a lot of happy discoveries, too.

"No wonder I love Vermont so much," he exulted as we hiked up Mount Sorak. Like Chris, he was fascinated by the well-guarded military bunkers: just like the G.I. Joes he used to play with, but this was real! And was it just coincidence that the decorative lily ponds we saw looked just like the one that he dug at home in our own front yard five summers ago? Yet the nights we stayed at traditional Korean inns and slept on thick, brightly colored floor mats, it was Josh who longed for a "real" bed. On our last morning at breakfast, Josh concluded: "I feel ties to this country. Shocking how much Korean culture actually sifts into my personality." He grinned. "Horrors! A foreigner more Korean than anybody else? Impossible." He talked about the mountains, the lily ponds, the military uniforms. "And don't forget amethysts! It just happens to be my birthstone and just happens to be an export of Korea.

"Still, I'm glad I didn't grow up in Korea. I can't stand spicy food!" Josh let us know that he was glad to be heading home to his *real* home, with us, in the United States.

End of story? No, not really. Josh spent the rest of the summer at home with us and we fought—a lot—over stupid things. He was outrageously messy. I was a nag. "Boy!" he said after one confrontation, "I'll be glad to get out of here and go to college!" "Good!" I retorted. "I'll be glad, too." In between, and privately, of course, I mourned his leaving, and recalling the old days, I suspected that he must have mixed feelings at the prospect of this major change in his life.

The day we helped Josh move into his freshman dorm, he seemed optimistic, confident and happy. We felt good, too, as he kissed us goodbye and we left on the five-hour trip back home. It was only last night on the phone, talking about this article, that Josh sheepishly told me a secret: Just before he left home that day, he took a very, *very,* very big breath of air and held it for as long as he could.

What Feels Like the World

BY

RICHARD

BAUSCH

~

Very early in the morning, too early, he hears her trying to jump rope out on the sidewalk below his bedroom window. He wakes to the sound of her shoes on the concrete, her breathless counting as she jumps—never more than three times in succession—and fails again to find the right rhythm, the proper spring in her legs to achieve the thing, to be a girl jumping rope. He gets up and moves to the window and, parting the curtain only slightly, peers out at her. For some reason he feels he must be stealthy, must not let her see him gazing at her from this window. He thinks of the heartless way children tease the imperfect among them, and then he closes the curtain.

She is his only granddaughter, the unfortunate inheritor of his big-boned genes, his tendency toward bulk, and she is on a self-induced program of exercise and dieting, to lose weight. This is in preparation for the last meeting of the PTA, during which children from the fifth and sixth grades will put on a gymnastics demonstration. There will be a vaulting horse and a minitrampoline, and everyone is to participate. She wants to be able to do at least as well as the other children in her class, and so she has been trying exercises to improve her coordination and lose the weight that keeps her rooted to the ground. For the past two weeks she has been eating only one meal a day, usually lunch, since that's the meal she eats at school, and swallowing cans of juice at other mealtimes. He's afraid of anorexia but trusts her calm determination to get ready for the event. There seems no desperation, none of the classic symptoms of the disease. Indeed, this project she's set for herself seems quite sane: to lose ten pounds, and to be able to get over the vaulting horse—in fact, she hopes that she'll be able to do a handstand on it and, curling her head and shoulders, flip over to stand upright on the other side. This, she has told him, is the outside hope. And in two weeks of very grown-up discipline and single-minded effort, that hope has mostly disappeared; she's still the only child in the fifth grade who has not even been able to propel herself over the horse, and this is the day of the event. She will have one last chance to practice at school today, and so she's up this early, out on the lawn, straining, pushing herself.

He dresses quickly and heads downstairs. The ritual in the mornings is simplified by the fact that neither of them is eating breakfast. He makes the orange juice, puts vitamins on a saucer for them both. When he glances out the living-room window, he sees that she is now doing somersaults in the dewy grass. She does three of them while he watches, and he isn't stealthy this time but stands in the window with what he hopes is an approving, unworried look on his face. After each somersault she pulls her sweat shirt down, takes a deep breath, and begins again, the arms coming down slowly, the head ducking slowly under; it's as if she falls on her back, sits up, and then stands up. Her cheeks are ruddy with effort. The moistness of the grass is on the sweat suit, and in the ends of her hair. It will rain this morning—there's thunder beyond the trees at the end of the street. He taps on the window, gestures, smiling, for her to come in. She waves at him, indicates that she wants him to watch her, so he watches her. He applauds when she's finished—three hard, slow tumbles. She claps her hands together as if to remove dust from them and comes trotting to the door. As she moves by him he tells her she's asking for a bad cold, letting herself get wet so early in the morning. It's his place to nag. Her glance at him acknowledges this.

"I can't get the rest of me to follow my head," she says about the somersaults.

They go into the kitchen and she sits down, pops a vitamin into her mouth, and takes a swallow of the orange juice. "I guess I'm not going to make it over that vaulting horse after all," she says suddenly.

"Sure you will."

"I don't care." She seems to pout. This is the first sign of true discouragement she's shown.

He's been waiting for it. "Brenda—honey, sometimes people aren't good at these things. I mean, I was never any good at it."

"I bet you were," she says. "I bet you're just saying that to make me feel better."

"No," he says, "really."

He's been keeping to the diet with her, though there have been times during the day when he's cheated. He no longer has a job, and the days are long; he's hungry all the time. He pretends to her that he's still going on to work in the mornings after he walks her to school, because he wants to keep her sense of the daily balance of things, of a predictable and orderly routine, intact. He believes this is the best way to deal with grief—simply to go on with things, to keep them as much as possible as they have always been. Being out of work doesn't worry him, really: he has enough money in savings to last awhile. At sixty-one, he's almost eligible for Social Security, and he gets monthly checks from the girl's father, who lives with another woman, and other children, in Oregon. The father has been very good about keeping up the payments, though he never visits or calls. Probably he thinks the money buys him the privilege of remaining aloof, now that Brenda's mother is gone. Brenda's mother used to say he was the type of man who learned early that there was nothing of substance anywhere in his soul, and spent the rest of his life trying to hide this fact from himself. No one was more upright, she would say, no one more honorable, and God help you if you ever had to live with him. Brenda's father was the subject of bitter sarcasm and scorn. And yet, perhaps not so surprisingly, Brenda's mother would call him in those months just after the divorce, when Brenda was still only a toddler, and she would try to get the baby to say things to him over the phone. And she would sit there with Brenda on her lap and cry after she had hung up.

"I had a doughnut yesterday at school," Brenda says now.

"That's lunch. You're supposed to eat lunch."

"I had spaghetti, too. And three pieces of garlic bread. And pie. And a big salad."

"What's one doughnut?"

"Well, and I didn't eat anything the rest of the day."

"I know," her grandfather says. "See?"

They sit quiet for a little while. Sometimes they're shy with each other—more so lately. They're used to the absence of her mother by now—it's been almost a year—but they still find themselves missing a beat now and then, like a heart with a valve almost closed. She swallows the last of her juice and then gets up and moves to the living room, to stand gazing out at the yard. Big drops have begun to fall. It's a storm, with rising wind and, now, very loud thunder. Lightning branches across the sky, and the trees in the yard disappear in sheets of rain. He has come to her side, and he pretends an interest in the details of the weather, remarking on the heaviness of the rain, the strength of the wind. "Some storm," he says finally. "I'm glad we're not out in it." He wishes he could tell what she's thinking, where the pain is; he wishes he could be certain of the harmlessness of his every word. "Honey," he ventures, "we could play hooky today. If you want to."

"Don't you think I can do it?" she says.

"I know you can."

She stares at him a moment and then looks away, out at the storm.

"It's terrible out there, isn't it?" he says. "Look at that lightning."

"You don't think I can do it," she says.

"No. I know you can. Really."

"Well, I probably can't."

"Even if you can't. Lots of people—lots of people never do anything like that."

"I'm the only one who can't that I know."

"Well, there's lots of people. The whole thing is silly, Brenda. A year from now it won't mean anything at all—you'll see."

She says nothing.

"Is there some pressure at school to do it?"

"No." Her tone is simple, matter-of-fact, and she looks directly at him.

"You're sure?"

She's sure. And of course, he realizes, there is pressure; there's the pressure of being one among other children, and being the only one among them who can't do a thing.

"Honey," he says lamely, "it's not that important."

When she looks at him this time, he sees something scarily unchildlike in her expression, some perplexity that she seems to pull down into herself. "It is too important," she says.

He drives her to school. The rain is still being blown along the street and above the low roofs of the houses. By the time they arrive, no more than five minutes from the house, it has begun to let up.

"If it's completely stopped after school," she says, "can we walk home?"

"Of course," he says. "Why wouldn't we?"

She gives him a quick wet kiss on the cheek. "Bye, Pops."

He knows she doesn't like it when he waits for her to get inside, and still he hesitates. There's always the apprehension that he'll look away or drive off just as she thinks of something she needs from him or that she'll wave to him and he won't see her. So he sits here with the car engine idling, and she walks quickly up and into the building. In the few seconds before the door swings shut, she turns and gives him a wave, and he waves back. The door is closed now. Slowly he lets the car glide forward, still watching the door. Then he's down the driveway, and he heads back to the house.

It's hard to decide what to do with his time. Mostly he stays in the house, watches television, reads the newspapers. there are household tasks, but he can't do anything she might notice, since he's supposed to be at work during these hours. Sometimes, just to please himself, he drives over to the bank and visits with his old co-workers, though there doesn't seem to be much to talk about anymore and he senses that he makes them all uneasy. Today he lies down on the sofa in the living room and rests awhile. At the windows the sun begins to show, and he thinks of driving into town, perhaps stopping somewhere to eat a light breakfast. He accuses himself with the thought and then gets up and turns on the television. There isn't anything of interest to watch, but he watches anyway. The sun is bright now out on the lawn, and the wind is the same, gusting and shaking the window frames. On television he sees feasts of incredible sumptuousness, almost nauseating in the impossible brightness and succulence of the food: advertisements from cheese companies, dairy associations, the makers of cookies and pizza, the sellers of seafood and steaks. He's angry with himself for waiting to cheat on the diet. He thinks of Brenda at school, thinks of crowds of children, and it comes to him more painfully than ever that he can't protect her. Not any more than he could ever protect her mother.

He goes outside and walks up the drying sidewalk to the end of the block. The sun has already dried most of the morning's rain, and the wind is warm. In the sky are great stormy Matterhorns of cumulus and wide patches of the deepest blue. It's a beautiful day, and he decides to walk over to the school. Nothing in him voices this decision; he simply begins to walk. He knows without having to think about it that he can't allow her to see him yet he feels compelled to take the risk that she might; he feels a helpless wish to watch over her, and, beyond this, he entertains the vague notion that by seeing her in her world he might be better able to be what she needs in his.

So he walks the four blocks to the school and stands just beyond the playground, in a group of shading maples that whisper and sigh in the wind. The playground is empty. A bell rings somewhere in the building, but no one comes out. It's not even eleven o'clock in the morning. He's too late for morning recess and too early for the afternoon one. He feels as though she watches him to make his way back down the street.

His neighbor, Mrs. Eberhard, comes over for lunch. It's a thing they planned, and he's forgotten about it. She knocks on the door, and when he opens it she smiles and says, "I knew you'd forget." She's on a diet too, and is carrying what they'll eat: two apples, some celery and carrots. It's all in a clear plastic bag, and she holds it toward him in the palms of her hands as though it were piping hot from the oven. Jane Eberhard is relatively new in the neighborhood. When Brenda's mother died, Jane offered to cook meals and regulate things, and for a while she was like another member of the family. She's moved into their lives now, and sometimes they all forget the circumstances under which the friendship began. She's a solid, large-hipped woman of fifty-eight, with clear, young blue eyes and gray hair. The thing she's good at is sympathy; there's something oddly unspecific about it, as if it were a beam she simply radiates.

"You look so worried," she says now, "I think you should be proud of her."

They're sitting in the living room, with the plastic bag on the coffee table before them. She's eating a stick of celery.

"I've never seen a child that age put such demands on herself," she says.

"I don't know what it's going to do to her if she doesn't make it over the damn thing," he says.

"It'll disappoint her. But she'll get over it."

"I don't guess you can make it tonight."

"Can't," she says. "Really. I promised my mother I'd take her to the ocean this weekend. I have to go pick her up tonight."

"I walked over to the school a little while ago."

"Are you sure you're not putting more into this than she is?"

"She was up at dawn this morning, Jane. Didn't you see her?"

Mrs. Eberhard nods. "I saw her."

"Well?" he says.

She pats his wrist. "I'm sure it won't matter a month from now."

"No," he says, "that's not true. I mean, I wish I could believe you. But I've never seen a kid work so hard."

"Maybe she'll make it."

"Yes," he says. "Maybe."

Mrs. Eberhard sits considering for a moment, tapping the stick of celery against her lower lip. "You think it's tied to the accident in some way, don't you?"

"I don't know," he says, standing, moving across the room. "I can't get through somehow. It's been all this time and I still don't know. She keeps it all to herself—all of it. All I can do is try to be there when she wants me to be there. I don't know—I don't even know what to say to her."

"You're doing all you can do, then."

"Her mother and I..." he begins. "She—we never got along that well."

"You can't worry about that now."

Mrs. Eberhard's advice is always the kind of practical good advice that's impossible to follow.

He comes back to the sofa and tries to eat one of the apples, but his appetite is gone. This seems ironic to him. "I'm not hungry now," he says.

"Sometimes worry is the best thing for a diet."

"I've always worried. It never did me any good, but I worried."

"I'll tell you," Mrs. Eberhard says. "It's a terrific misfortune to have to be raised by a human being."

He doesn't feel like listening to this sort of thing, so he asks her about her husband, who is with the government in some capacity that requires him to be both secretive and mobile. He's always off to one country or another, and this week he's in India. It's strange to think of someone traveling as much as he does without getting hurt or killed.

Mrs. Eberhard says she's so used to his being gone all the time that next year, when he retires, it'll take a while to get used to having him underfoot. In fact, he's not a very likable man; there's something murky and unpleasant about him. The one time Mrs. Eberhard brought him to visit, he sat in the living room and seemed to regard everyone with detached curiosity, as if they were all specimens on a dish under a lens. Brenda's grandfather had invited some old friends over from the bank—everyone was being careful not to let on that he wasn't still going there every day. It was an awkward two hours, and Mrs. Eberhard's husband sat with his hands folded over his rounded belly, his eyebrows arched. When he spoke, his voice was cultivated and quiet, full of self-satisfaction and haughtiness. They had been speaking in low tones about how Jane Eberhard had moved in to take over after the accident, and Mrs. Eberhard's husband cleared his throat, held his fist gingerly to his mouth, pursed his lips, and began a soft-spoken, lecturelike monologue about his belief that there's no such thing as an accident. His considered opinion was that there are subconscious explanations for everything. Apparently, he thought he was entertaining everyone. He sat with one leg crossed over the other and held forth in his calm, magisterial voice, explaining how every-thing can be reduced to a matter of conscious or subconscious will. Finally his wife asked him to let it alone, please, drop the subject.

"For example," he went on, "there are many collisions on the high-way in which no one appears to have applied brakes before impact, as if something in the victims had decided on death. And of course there are the well-known cases of people stopped on railroad tracks, with plenty of time to get off, who simply do not move. Perhaps it isn't being frozen by the perception of one's fate but a matter of decision making, of will. The victim decides on his fate."

"I think we've had enough, now," Jane Eberhard said.

The inappropriateness of what he had said seemed to dawn on him then. He shifted in his seat and grew very quiet, and when the evening was over he took Brenda's grandfather by the elbow and apologized. But even in the apology there seemed to be a species of condescension, as if he were really only sorry for the harsh truth of what he had wrongly deemed it necessary to say. When everyone was gone, Brenda said, "I don't like that man."

"Is it because of what he said about accidents?" her grandfather asked.

She shook her head. "I just don't like him."

"It's not true, what he said, honey. An accident is an accident."

She said, "I know." But she would not return his gaze.

"Your mother wasn't very happy here, but she didn't want to leave us. Not even—you know, without...without knowing it or anything."

"He wears perfume," she said, still not looking at him.

"It's cologne. Yes, he does—too much of it."

"It smells," she said.

In the afternoon he walks over to the school. The sidewalks are crowded with children, and they all seem to recognize him. They carry their books and papers and their hair is windblown and they run and wrestle with each other in the yards. The sun's high and very hot, and most of the clouds have broken apart and scattered. There's still a fairly steady wind, but it's gentler now, and there's no coolness in it.

Brenda is standing at the first crossing street down the hill from the school. She's surrounded by other children yet seems separate from them somehow. She sees him and smiles. He waits on his side of the intersection for her to cross, and when she reaches him he's careful not to show any obvious affection, knowing it embarrasses her.

"How was your day?" he begins.

"Mr. Clayton tried to make me quit today."

He waits.

"I didn't get over," she says. "I didn't even get close."

"What did Mr. Clayton say?"

"Oh—you know. That it's not important. That kind of stuff."

"Well," he says gently, "*is* it so important?"

"I don't know." She kicks at something in the grass along the edge of the sidewalk—a piece of a pencil someone else had discarded. She bends, picks it up, examines it, and then drops it. This is exactly the kind of slow, daydreaming behavior that used to make him angry and impatient with her mother. They walk on. She's concentrating on the sidewalk before them, and they walk almost in step.

"I'm sure I could never do a thing like going over a vaulting horse when I was in school," he says.

"Did they have that when you were in school?"

He smiles. "It was hard getting everything into the caves. But sure, we had that sort of thing. We were an advanced tribe. We had fire, too."

"Okay," she's saying, "okay, okay."

"Actually, with me, it was pull-ups. We all had to do pull-ups. And I just couldn't do them. I don't think I ever accomplished a single one in my life."

"I can't do pull-ups," she says.

"They're hard to do."

"Everybody in the fifth and sixth grades can get over the vaulting horse," she says.

How much she reminds him of her mother. There's a certain mobility in her face, a certain willingness to assert herself in the smallest gesture of the eyes and mouth. She has her mother's green eyes, and now he tells her this. He's decided to try this. He's standing, quite shy, in her doorway, feeling like an intruder. She's sitting on the floor, one leg outstretched, the other bent at the knee. She tries to touch her forehead to the knee of the outstretched leg, straining, and he looks away.

"You know?" he says. "They're just the same color—just that shade of green."

"What was my grandmother like?" she asks, still straining.

"She was a lot like your mother."

"I'm never going to get married."

"Of course you will. Well, I mean—if you want to, you will."

"How come you didn't ever get married again?"

"Oh," he says, "I had a daughter to raise, you know."

She changes position, tries to touch her forehead to the other knee.

"I'll tell you, that mother of yours was enough to keep me busy. I mean, I called her double trouble, you know, because I always said she was double the trouble a son would have been. That was a regular joke around here."

"Mom was skinny and pretty."

He says nothing.

"Am I double trouble?"

"No," he says.

"Is that really why you never got married again?"

"Well, no one would have me, either."

"Mom said you liked it."

"Liked what?"

"Being a widow."

"Yes, well," he says.

"Did you?"

"All these questions," he says.

"Do you think about Grandmom a lot?"

"Yes," he says. "That's—you know, we remember our loved ones."

She stands and tries to touch her toes without bending her legs. "Sometimes I dream that Mom's yelling at you and you're yelling back."

"Oh, well," he says, hearing himself say it, feeling himself back down from something. "That's—that's just a dream. You know, it's nothing to think about at all. People who love each other don't agree sometimes—it's—it's nothing. And I'll bet these exercises are going to do the trick."

"I'm very smart, aren't I?"

He feels sick, very deep down. "You're the smartest little girl I ever saw."

"You don't have to come tonight if you don't want to," she says. "You can drop me off if you want, and come get me when it's over."

"Why would I do that?"

She mutters. "*I* would."

"Then why don't we skip it?"

"Lot of good *that* would do," she says.

For dinner they drink apple juice, and he gets her to eat two slices of dry toast. The apple juice is for energy. She drinks it slowly and then goes into her room to lie down, to conserve her strength. She uses the word *conserve*, and he tells her he's so proud of her vocabulary. She thanks him. While she rests, he does a few household chores, trying really just to keep busy. The week's newspapers have been piling up on the coffee table in the living room, the carpets need to be vacuumed, and the whole house needs dusting. None of it takes long enough; none of it quite distracts him. For a while he sits in the living room with a newspaper in his lap and pretends to be reading it. She's restless too. She comes back through to the kitchen, drinks another glass of apple juice, and then joins him in the living room, turns the television on. The news is full of traffic deaths, and she turns to one of the local stations that shows reruns of old situation comedies. They both watch *M*A*S*H* without really taking it in. She bites the cuticles of her nails, and her gaze wanders around the room. It comes to him that he could speak to her now, could make his way through to her grief—and yet he knows that he will do no such thing; he can't even bring himself to speak at all. There are regions of his own sorrow that he simply lacks the strength to explore, and so he sits there watching her restlessness, and at last it's time to go over to the school. Jane Eberhard makes a surprise visit, bearing a handsome good-luck card she's fashioned herself. She kisses Brenda, behaves exactly as if Brenda were going off to some dangerous, faraway place. She stands in the street and waves at them as they pull away, and Brenda leans out the window to shout goodbye. A moment later, sitting back and staring out at the dusky light, she says she feels a surge of energy, and he tells her she's way ahead of all the others in her class, knowing words like *conserve* and *surge.*

"I've always known them," she says.

It's beginning to rain again. Clouds have been rolling in from the east, and the wind shakes the trees. Lightning flickers on the other side of the clouds. Everything seems threatening, relentless. He slows down. There are many cars parked along both sides of the street. "Quite a turnout," he manages.

"Don't worry," she tells him brightly. "I still feel my surge of energy."

It begins to rain as they get out of the car, and he holds his sport coat like a cape to shield her from it. By the time they get to the open front doors, it's raining very hard. People are crowding into the cafeteria, which has been transformed into an arena for the event—chairs set up on four sides of the room as though for a wrestling match. In the center, at the end of the long bright-red mat, are the vaulting horse and the minitrampoline. The physical-education teacher, Mr. Clayton, stands at the entrance. He's tall, thin, scraggly-looking, a boy really, no older than twenty-five.

"There's Mr. Clayton," Brenda says.

"I see him."

"Hello, Mr. Clayton."

Mr. Clayton is quite distracted, and he nods quickly, leans toward Brenda, and points to a doorway across the hall. "Go on ahead," he says. Then he nods at her grandfather.

"This is it," Brenda says.

Her grandfather squeezes her shoulder, means to find the best thing to tell her, but in the next confusing minute he's lost her; she's gone among the others and he's being swept along with the crowd entering the cafeteria. He makes his way along the walls behind the chairs, where a few other people have already gathered and are standing. At the other end of the room a man is speaking from a lectern about old business, new officers for the fall. Brenda's grandfather recognizes some of the people in the crowd. A woman looks at him and nods, a familiar face he can't quite place. She turns to look at the speaker. She's holding a baby, and the baby's staring at him over her shoulder. A moment later, she steps back to stand beside him, hefting the baby higher and patting its bottom.

"What a crowd," she says.

He nods.

"It's not usually this crowded."

Again, he nods.

The baby protests, and he touches the miniature fingers of one hand—just a baby, he thinks, and everything still to go through.

"How is—um...Brenda?" she says.

"Oh," he says, "fine." And he remembers that she was Brenda's kindergarten teacher. She's heavier than she was then and her hair is darker. She has a baby now.

"I don't remember all my students," she says, shifting the baby to the other shoulder. "I've been home now for eighteen months, and I'll tell you, it's being at the PTA meeting that makes me see how much I *don't* miss teaching."

He smiles at her and nods again. He's beginning to feel awkward. The man is still speaking from the lectern, a meeting is going on, and this woman's voice is carrying beyond them, though she says everything out of the side of her mouth.

"I remember the way you used to walk Brenda to school every morning. Do you still walk her to school?"

"Yes."

"That's so nice."

He pretends an interest in what the speaker is saying.

"I always thought it was so nice to see how you two got along together—I mean these days it's really rare for the kids even to know who their grandparents *are*, much less have one to walk them to school in the morning. I always thought it was really something." She seems to watch the lectern for a moment, and then speaks to him again, this time in a near whisper. "I hope you won't take this the wrong way or anything, but I just wanted to say how sorry I was about your daughter. I saw it in the paper when Brenda's mother...well, you know, I just wanted to tell you how sorry. When I saw it in the paper, I thought of Brenda, and how you used to walk her to school. I lost my sister in an automobile accident, so I know how you feel—it's a terrible thing. Terrible. An awful thing to have happen. I mean it's much too sudden and final and everything. I'm afraid now every time I get into a car." She pauses, pats the baby's back, then takes something off its ear. "Anyway, I just wanted to say how sorry I was."

"You're very kind," he says.

"It seems so senseless," she murmurs. "There's something so senseless about it when it happens. My sister went through a stop sign. She just didn't see it, I guess. But it wasn't a busy road or anything. If she'd come along one second later or sooner nothing would've happened. So senseless. Two people driving two different cars coming along on two roads on a sunny afternoon and they come together like that. I mean—what're the chances, really?"

He doesn't say anything.

"How's Brenda handling it?"

"She's strong," he says.

"I would've said that," the woman tells him. "Sometimes I think the children take these things better than the adults do. I remember when she first came to my class. She told everyone in the first minute that she'd come from Oregon. That she was living with her grandfather, and her mother was divorced."

"She was a baby when the divorce—when she moved here from Oregon."

This seems to surprise the woman. "Really," she says, low. "I got the impression it was recent for her. I mean, you know, that she had just come from it all. It was all very vivid for her, I remember that."

"She was a baby," he says. It's almost as if he were insisting on it. He's heard this in his voice, and he wonders if she has, too.

"Well," she says, "I always had a special place for Brenda. I always thought she was very special. A very special little girl."

The PTA meeting is over, and Mr. Clayton is now standing at the far door with the first of the charges. They're all lining up outside the door, and Mr. Clayton walks to the microphone to announce the program. The demonstration will commence with the minitrampoline and the vaulting horse: a performance by the fifth- and sixth-graders. There will also be a break-dancing demonstration by the fourth-grade class.

"Here we go," the woman says. "My nephew's afraid of the minitramp."

"They shouldn't make them do these things," Brenda's grandfather says, with a passion that surprises him. He draws in a breath. "It's too hard," he says, loudly. He can't believe himself. "They shouldn't have to go through a thing like this."

"I don't know," she says vaguely, turning from him a little. He has drawn attention to himself. Others in the crowd are regarding him now—one, a man with a sparse red beard and wild red hair, looking at him with something he takes for agreement.

"It's too much," he says, still louder. "Too much to put on a child. There's just so much a child can take."

Someone asks gently for quiet.

The first child is running down the long mat to the minitrampoline; it's a girl, and she times her jump perfectly, soars over the horse. One by one, other children follow. Mr. Clayton and another man stand on either side of the horse and help those who go over on their hands. Two or three go over without any assistance at all, with remarkable effortlessness and grace.

"Well," Brenda's kindergarten teacher says, "there's my nephew."

The boy hits the minitramp and does a perfect forward flip in the air over the horse, landing upright and then rolling forward in a somersault.

"Yea, Jack!" she cheers. "No sweat! Yea, Jackie boy!"

The boy trots to the other end of the room and stands with the others; the crowd is applauding. The last of the sixth-graders goes over the horse, and Mr. Clayton says into the microphone that the fifth-graders are next. It's Brenda who's next. She stands in the doorway, her cheeks flushed, her legs looking too heavy in the tights. She's rocking back and forth on the balls of her feet, getting ready. It grows quiet. Her arms swing slightly, back and forth, and now, just for a moment, she's looking at the crowd, her face hiding whatever she's feeling. It's as if she were merely curious as to who is out there, but he knows she's looking for him, searching the crowd for her grandfather, who stands on his toes, unseen against the far wall, stands there thinking his heart might break, lifting his hand to wave.

Woman
With No
Face

~

BY

ALICE

LEE

ld Mother has spread her blanket of stars over all. My eyes are heavy with dreams as I sit beside Kookum. As I doze off, sitting beside my grandmother, I dream a quilt of blueberries wrapped around me. In my dream I nibble a corner of the quilt. I love the taste of blueberries.

Kookum shakes the dreams from my eyes. "The campfire is going out. Firedancer is ending his dance. Tomorrow we will pack up camp and return home. I will need your help to make pies and jams with the berries we've picked. Starlet, you must go to the river and bring water for morning."

I turn and look into my grandmother's face and know she sees the fear in my eyes but Kookum turns to the campfire in silence. "Please come with me," I plead silently. "It's dark. I don't know this dark. I wish I were back in the city with my mom. It's never this dark in the city."

I walk carefully toward the river. The path which is my friend in the day has become a stranger. My new running shoes shine white in the night. Kookum gave them to me. When I was smaller she used to make me beaded moccasins but now she says her hands are getting too old to sew and bead.

Many night sounds are loud in my ears. Fear begins a song in my throat. Fear's voice jumps from my throat and sings in my head. The voice becomes one with the darkness.

River is still while dreaming. As I lower my jug, my face becomes part of River's dream, singing now with fear on reflection under. River is still once more.

Another face climbs slowly up from River's depths. Tangled brown weed hair falls over his face. Pieces of garbage are caught in the tangles. My heart is still as River stirs in sleep and washes the hair away from the face. A woman with no face looks back at me.

The dark smell of death is strong in my nostrils. An evil smell. Fear beats in my heart as I drop my jug and turn to run. The earthen jug lies empty, forgotten. Behind me, it is as though an open throat reaches out to eat my flesh. I'm chased by the shadow of Fear's song as I run back to my grandmother.

My breathing is fast and loud as I run close to Kookum. She sits as still as the fire laying on the earth bed at her feet. "I saw a woman with no face!" I point to the river.

Kookum reaches out and holds me close to her. "The earth is a gift. Gifts must not be wasted. There were those people who held sacred all that was of the earth. There was respect for all things living and dead. Peoples' ways have changed. The earth is hurting. Her body is scarred and bruised. This woman with no face has shown herself to you. She is the injured spirit of the earth. We have to know that to hurt the earth is to hurt ourselves."

"I've become an old one and will soon die. I'll return to the earth and comfort her with my bones. Starlet, it's your birthright to hold sacred all that is of the earth. This woman with no face is of the earth. I'm going to give you a song in your mother's tongue. In our language. This is a holy song, a healing song. You must share it with others so that they too can know how to save the earth."

Kookum begins to sing. I listen to her voice. I know somehow that they are cries and chants of another time. The voice of a people. I listen so that I will know the words. I listen so that I too will sing for others to listen.

Hero

BY

SHEREE

FITCH

"We are going to the jungle"
my father and my son inform me

"Be careful of the lions"
I tell them

My son ties a red terrycloth
superman cape around his neck
my father takes a walking stick
dog on leash
the trio
set off on safari

They are gone many days and nights
or so it seems
for I do worry about the lions
when my cubs have wandered off without me

No need to worry
here they come now

My father is wearing the superman cape
pretending to fly through the neighbourhood
shouting: superman! superman!

My son is running by his side
the dog is yapping
my father in fluorescent red
is making a spectacle of himself

I can see that to my son
this is not a game
of just pretend

My father scoops him up

I watch as they lift into the air
And fly the rest of the way home.

Do Not Go Gentle Into That Good Night

BY

DYLAN

THOMAS

Do not go gentle into that good night,
Old age should burn and rave at close of day;
Rage, rage against the dying of the light.

Though wise men at their end know dark is right,
Because their words had forked no lightning they
Do not go gentle into that good night.

Good men, the last wave by, crying how bright
Their frail deeds might have danced in a green bay,
Rage, rage against the dying of the light.

Wild men who caught and sang the sun in flight,
And learn, too late, they grieved it on its way,
Do not go gentle into that good night.

Grave men, near death, who see with blinding sight
Blind eyes could blaze like meteors and be gay,
Rage, rage against the dying of the light.

And you, my father, there on the sad height,
Curse, bless, me now with your fierce tears, I pray.
Do not go gentle into that good night.
Rage, rage against the dying of the light.

Power Failure

~

BY

JANE

RULE

Laura Thornstrom was watching the late news, a habit which often gave her nightmares but consoled her that she hadn't much longer to live, when the power failed. The color fled from the screen like water going down the drain, and she heard the silencing of all the small hummings in the house, fridge, freezer, pump. In the dark she heard only her heart beating and wondered at its independence from all those other gadgets she needed to stay alive. Every time this winter the power had failed, she tested the notion that she was too old to stay on alone on this little island so vulnerable to the weather and isolated enough to wait sometimes several days for help. In a power failure she had to haul not only wood for heat and cooking but water from the rain barrels, if they weren't frozen, for washing and flushing the toilet. Even the thought wore her out.

She reached for the matches on the table beside her and lighted a candle. With its light she located one of the batteried fluorescent lights, safer than any of the old fashioned lanterns now that she was both inclined to stumble and to be absentminded. She would do nothing tonight but go to bed, getting her extra quilt out to make up for the failure of her electric blanket. There was no wind; so perhaps there would be power in the morning.

The weight of the quilt on her arthritic feet made her moan, a bad habit born of living alone, for the sound of it didn't comfort her, only made her feel sorrier for herself. But what did it matter when there was no one else around to be troubled by her moods? In the two years since Thorny had died, Laura had retreated not into a second childhood but into a second adolescence. Often sulky, melodramatic, and clumsy, she wondered if the only incentive to be an adult was an intimate audience before whom one was ashamed to brood and complain. Certainly, it was only since Thorny had died that she'd been angry with him in that hopeless way of adolescents, so often angry about what couldn't be helped, shouting, "It isn't fair!"

Well, it wasn't fair that the power was out and her feet hurt and her husband was dead. She didn't complain ever to her children because they would only say she was mad to stay here and should move back to the mainland where among them they could look after her.

She and Thorny hadn't had enough money to retire in the city. They didn't care. They sold their city house and fixed up their summer place, insulating it, getting a good wood stove, and putting in electric heaters for when they were too old to chop and haul wood. It had more bedrooms than they needed, but that had assured them visits from children and grandchildren, occasions Laura now anticipated with a mixture of pleasure and dread. She hadn't really the energy any more to cook for them all and cope with the boisterous energy of the little ones. But, if she admitted that to anyone, the pressure was on again for her to leave the island.

They always came in the spring, summer, and fall when Laura loved the island and knew she would never leave it. She wore herself out before they even arrived, making sure no daughter or daughter-in-law would find the oven neglected, the kitchen floor sticky, the spare rooms fit only for spiders which were Laura's undisturbed companions when she was alone. Her sons and sons-in-law were less insistent with her, helped instead by seeing to it she had plenty of wood to last the winter, stacked conveniently near the house, checking gutters, floor boards, light switches. Nobody expected her to take over Thorny's jobs to prove she could stay on alone. Sometimes she wanted to snap at the girls, tell them she didn't have to keep an AA motel rating for half a dozen guests to be permitted an independent old age. But she knew temper, too, was one of the signs.

Turning painfully in her bed, she tried to stop listening to all those voices, to the absolute silence of the house. Had she remembered to turn off everything, lights, heaters, the electric stove, the television so that the house wouldn't blaze her into waking if the power came on before morning? It wouldn't. And she didn't care if it did. It was simply stupid to worry about being wakened when she couldn't go to sleep.

It snowed in the night. Laura could tell by the white light in her bedroom when she opened her eyes. She could tell by the coldness of her nose that the power was still off. She felt entombed in her bed, the heavy quilt weighing down on her stiff, painful joints. Then she thought of the birds. In the snow they must be fed, particularly her pair of variegated thrushes who stayed with her through the winter, not like those rascal aristocrats who, like her closest human neighbors, went south for the winter. She and Thorny had, too, for a month or so, but they were always glad to get home.

Laura groaned as she hoisted herself out of bed, this morning so slowly that she knew one of these mornings she just wouldn't make it. She had just finished her slow and layered dressing when there was a knocking at her door.

"Come in!" she shouted, knowing it would take her painful minutes to get to a door she never locked.

"It's me, Jimmy," called the voice in uncertain, deep register, "come to start your wood stove."

"Why aren't you in school?" she asked, having made it to her bedroom door which opened onto the living room.

"It's Saturday, Mrs. Thornstrom," Jimmy said to her, balancing an armload of kindling and logs. "I already swept your path, but I'll bring in enough wood for the day."

"That's very kind of you, Jimmy."

"My dad says there's no use in having ten kids if they're all good for nothing," Jimmy answered with a grin. "If the power's still off tonight, Peter's going to stop by. He's not worth much unless you tell him."

"He's only seven, Jimmy," Laura reminded him.

"Yeah, but his memory's as bad as old Mr. Apple's. Senile at seven!"

Laura knew there was no point in offering to pay Jimmy. The O'Hea children were raised to do favors for people just like their father who managed to put enough food on the table because they all dressed out of the thrift shop.

"Can I make you a cup of cocoa?" Laura offered. "Think maybe there's a doughnut around somewhere, too."

"Don't mind if I do," Jimmy said. "I'll just get you some water."

"I'm afraid the barrels will be frozen."

"Nope. I checked. I'll bring enough in to put some in the tub."

As the kitchen warmed, Laura moved around more easily. She found the doughnuts, some stale bread and seeds for the birds, and, when she and Jimmy sat down together, she wondered why she had fretted in the night. A power failure, snow, anything of that sort turned an ordinary day into a holiday. What boy in the city would come pounding at her door ready to do all her chores for nothing but a snack which he stayed for more to keep her company than because he wanted it?

Laura knew not to ask about school. Jimmy was at an age when he fretted to be out of the classroom and into the woods or onto a fishing boat. His father would have let him, but his mother was determined that all the kids were going to finish high school. They spoke instead about the logging truck that had gone off the road last week, three fires in as many days so that the volunteer firemen all finally went home to bed and said either the island would burn down or the kids and women would have to cope with the next one.

Once Jimmy had gone and the birds were fed, Laura was briefly at loose ends without the chores she had expected to do for herself. Then she realized that on such a day she could simply settle to read while she had the light to. There was not even any point in cleaning up Jimmy's muddy boot prints since Peter's would be there this evening. A power failure was even a sort of luxury, if it didn't go on too long.

By evening, however, when her eyes were too tired to read by poor light, when there was no television, and Peter had so overloaded the stove that she would have to wait for hours to bank it, she decided to take her battery radio to bed, feeling the same kind of discouragement she had the night before. "If I should die before I wake" had seemed to her a morbid prayer when she was a child. Now she understood it as a prayer for the old, not morbid at all but simply mortal.

Static on the radio woke her, but, when she reached to turn it off, it was not on. Fire. Something was on fire! For no more than a second, those reluctant old bones held her prisoner of a prayer. Then she moved, no longer aware of pain. It was the kitchen, somehow the stove... The kitchen door, like a great hearth cast huge heat and light into the living room which had begun to fill with smoke. Laura backed away from it toward her front door, stepped into her boots, grabbed a coat and flashlight and stepped out into the shocking cold.

It woke her but left her inside the nightmare of fire, where she now must think what to do. It had already been to late to grab the fire extinguisher, too late to phone the fire department, too late to think what to save. She moved slowly, stupidly down the path and then turned back to believe what was happening. The fire had eaten into the second floor and flowered onto the roof. Even if the firemen arrived—and how could they when they didn't know?—it was already too late to save the house.

Laura watched the upstairs corner bedroom fold in on itself. She thought of the closet full of Thorny's clothes which she'd never got round to giving away and found herself weeping for them, the things of his life like his life itself gone. How perfectly silly she was, in tears over his clothes when she had nothing but the nightgown, coat and boots she stood in, *nothing.*

"Not so much as my handbag," she said aloud.

Then she saw flames dance into the cedar next to the house and realized the danger as well as the loss. She did not know if she'd been standing there a minute or half an hour. She must get help at once. Her nightgown fluttering in the snow around her boots, she lighted her way to the empty house of her neighbors. Fortunately she had left their key under a flower pot on their deck, fearing that otherwise, in her absent-mindedness, she would misplace it.

The fire number was stickered to their phone as it was in every island house. Once Laura had made the call, she sat in the dark house waiting to hear the fire station siren and then realizing it could not sound without power. Everyone would have to be phoned. What would they do if the woods were ablaze by the time they got there? Should she go back and try to attach the hose? The well pump was useless. In any case, Laura was too cold and weak to move. Should she call her son?

Laura didn't want any of her children to know, but they would have to know. She hadn't even a handkerchief to blow her nose. It made her feel guilty, sitting there sniffing, and then resentful, bitterly resentful to be cold and alone in the middle of the night. How she hated the night which had become like a personal enemy to her!

It must have been the overloaded stove...or a fire in the chimney. Well, what difference did any of that make now? She thought of her books from the Open Shelf Library. She would have to pay to replace those. She'd have to phone the bank for another cheque book. Things, after all, could be replaced. They were insured. Even the house was insured. But surrounding these reasonable thoughts, threatening to engulf them was the darkness.

Laura heard the wailing siren of the fire truck. Then its lights swung into view on the lane that passed this house on its way to hers. Behind the yellow fire engine came the tanker truck, and behind that a line of cars. Only when they had all gone by did Laura think she should let them know where she was.

She got as far as the deck and could go no further. Her stiff old bones simply refused to carry her back to that roaring, collapsing house. Then a late, lone car came along, and she signaled it with her flashlight.

It was Jim O'Hea with his three oldest boys.

"Run for it!" he ordered them.

Then he got out of the car, bounded up onto the deck and surrounded her with a strong arm.

"Come on," he said gently, "I'll take you home to Mum," as he always referred to his wife.

Laura did not protest. The little will she had had left her. In most of the houses they passed she could see the dementing darting of fire, and it frightened her newly each time, as if the whole neighborhood were kindling for disaster. In several she also saw random arcs of flashlights.

"The damned power!" Jim muttered. "We ought none of us to pay our bills this month. They should have had a crew here today."

"Well, at least I won't have to worry about the food in my freezer," Laura commented cheerfully, her social self functioning out there beyond the shuddering anxiety.

"You don't have to worry about a thing now," Jim said gently.

Even the younger O'Hea children were up, their mother urging them to finish glasses of milk so that she could herd them back to bed again.

The large room in which they cooked, ate, and entertained themselves was amply warmed by the wood stove, and it was cheerfully lighted by kerosene lamps. Laura gratefully accepted an old chair by the stove and the offer of a cup of tea.

"I sent over to Lyvia's for some clothes for you," Kathleen O'Hea said. "She's about your size. She ought to be here any minute."

"It's the middle of the night," Laura protested.

"About five in the morning by now, not much more than an hour before she'd be up anyway, and the engine's woke everybody anyway. Phone's been ringing off its ear."

"Oh dear," Laura said. "Oh dear."

"Burned right down, is it?" Kathleen asked.

"I suppose so. I waited at my neighbors."

"Sensible."

"I hope they can save the woods at the back."

"They'll do the best they can. How did it start, do you know?"

"In the chimney probably," Laura said, realizing that she didn't want little Peter implicated, though she was sure Jimmy would think to suspect him.

"Have you phoned your kids?"

"Not yet," Laura said. "I'll wait until they're up."

"One of them will want to be on the morning ferry."

"Oh, I don't want to go over there!" Laura said.

"Have to go some place," Kathleen commented reasonably.

Lyvia Tey was at the door with a large, cardboard box full of clothes.

"My poor dear!" she exclaimed. "Are you all right?"

Then followed the same questions Kathleen had asked before Lyvia got around to the box.

"I didn't know about shoes," Lyvia said. "Whether they'd fit you, but I brought some."

"Oh, my boots should do until I can get something else."

"It's a good thing I remembered underwear! Nothing but your nightgown?"

"Why don't you go into our bedroom and try them on?" Kathleen suggested.

Once alone in the bedroom with its large, unmade bed, Laura wanted simply to crawl into it, turn out the lamp and fall asleep, leaving all the horror and confusion and kindness behind her. She didn't want to put on Lyvia's clothes. She didn't like Lyvia's clothes. Her snobbish ingratitude shocked and oddly reassured her. At least her own taste hadn't burned up in the fire. Of course, she had to get dressed. She couldn't go on through the day in her nightgown, coat and boots, and day was coming.

Laura looked through the contents of the box, holding up a dress, then a pair of slacks, a blouse, a sweater, examining them in the shadowy light. It touched and shamed her that Lyvia hadn't brought over anything but her very best. Laura found a vest and underpants, discarded a bra, found a pair of socks. Then she didn't try on the shoes. There was only a small cracked mirror on a table, in which Laura could dimly see only her uncombed hair and unmade face. She hadn't a comb or lipstick to her name, and she looked simply awful. Thank heaven she had most of her own teeth, for no one could see that her bridge, soaking in the bathroom, had also been lost.

"How are you coming along in there?" Kathleen called. "May I come in?"

She brought with her a pitcher of warm water and a bowl, covered with a towel on which there was a comb and lipstick.

"I wish I could offer you a nice, hot bath," Kathleen said.

"Oh, that's just wonderful. I look such a fright!"

"You're all right. That's the main thing."

Other women had arrived by the time Laura reappeared. On the family's dining table were casseroles, bowls of salad, plates of cookies, pies, dozens of eggs, a ham, and in boxes piling up by the door were more clothes, bedding, towels, even a little battery radio.

"Oh, I can't accept all this!" Laura protested in dismay. "I don't even have anywhere to put it."

"You will," they reassured her.

She shouldn't have been surprised. In her years on the island she, too, had hurried to whatever place sheltered a burned out family, fire so terrifyingly common a disaster, taking food and clothing, household goods, and, if there was no insurance, money. It simply had never occurred to her that it would ever happen to her. Whatever was given she was bound to accept; yet it felt a kind of madness to her, things piling up around her when she had no place to go.

The children were getting up again as dawn light began to take over the shadows. Kathleen, helped by the other women, was getting breakfast ready. She sent one of the older girls on her bicycle to tell the firemen there would be breakfast ready for them when they were done.

Laura couldn't eat. She accepted another cup of tea and said, "I suppose I must call my son."

"My fishing tackle, too?" he asked, incredulous, and then took himself in hand.

It was only Thorny's clothes she had thought to mourn, not the closets full of belongings of her children and grandchildren, about which she hadn't concerned herself since her own were grown. She had no idea how many pairs of riding boots, tennis rackets, fishing rods, skin diving suits, to say nothing of ordinary clothing had been lost in the fire. She could only assure her son that it was everything. "Even my bridge," she added wryly.

"Your what?"

"My bridge: my teeth."

It was so ridiculous a conversation that Laura began to laugh, which, far from reassuring her son, convinced him she was hysterical. He promised to be on the morning boat.

"What day of the week is it?" Laura asked as she turned away from the phone.

"Sunday," Peter said, and then he leaned up against her, offering the burden of his small weight to comfort her. It did.

Kathleen's daughter came breathlessly back into the house, her presence commanding everyone's attention. It occurred to Laura that anyone from so large a family would have no self-consciousness in public speaking since even asking for the butter had to be done before a large audience.

"They said to say they were sorry, Mrs. Thornstrom, but it was pretty well gone before they got there."

Laura nodded. It was no news to her.

"But they've got the woods pretty well out, and they're just coming in for breakfast, except for Jimmy and one other who have to stay behind to watch."

The children were urged to finish breakfast quickly to give up their places to a dozen soot blackened and tired men who came discouragedly into the house, shaking their heads, cursing the power company which had so delayed their response, telling Laura how damned sorry they were, but there was nothing much left but the chimney and the bathtub. She had seen enough island fires to know what it would look like. She would not, as some did, try to sift through the ashes in hope of finding, oh, anything.

"Shut up a minute, all of you," Jim O'Hea commanded. "Can you hear that?"

It was the hum of the refrigerator. The power was back on.

"Thank God!" Kathleen said. "You can all wash."

One of the kids went immediately to the television and turned it on. Another tried all the by now unnecessary light switches. The women urged the men off to the bathroom to leave the kitchen sink free for washing dishes for the next round of breakfast.

All through the morning, more people arrived, bringing food and gifts. Only much later Laura also found in a handbag several hundred dollars in cash. She was offered places to stay, advice about rebuilding, "Hell of a way to get a new house, but it makes nice work for some of us."

By the time Thorny junior arrived, his stricken face looked out of place in the cheerful crowd. But Laura was very glad to see him. Maybe he could figure out what to do with the carloads of goods turning the O'Hea's into a warehouse.

"Now that you're here, darling, I think I'll just lie down for a while."

Kathleen led her back into the bedroom which she had somehow managed to straighten. Laura lay down and slept for a few minutes or an hour, waking to escape the fire that burned all around her, hoping, until she realized where she was, that it was only her dream that had frightened her.

Her room at Thorny junior's house was no more reassuring as she woke again and again from dreams of fire. She wept for ridiculous things like her own needle and thimble, her little travel clock, and was unconsoled by the swiftness with which her children replaced them. She was furious with herself that such irrationalities only made them more convinced that the fire was a blessing in disguise to bring her back to what sense she had left.

"Think of this as your house now, Mother," Thorny junior said.

"But it isn't," she protested. "It's yours."

If only she could get one good night's sleep, she could begin to pull herself together.

"You mustn't keep waiting on me like this," she said to her daughter-in-law as she came in with the now accustomed breakfast tray.

"It's no trouble, Mother. At your age, you've got a right to start the day slowly."

"Most days never get started at all," Laura observed.

"You just rest."

Resting gave Laura time to grieve for her house, yet what had it been but a place too big for her with too much of her own and other people's clutter? The memories, well, they hadn't burned after all. She didn't need the house to go on being mad at Thorny for being dead.

It was the insurance release papers that got her out of bed.

"No, I'm not signing them, son."

"But why not?"

"Because I'm going to rebuild."

"What?"

"I only just realized it," Laura said. "But a small house, one my own size."

"Where?"

"Why, right where the other one was. And don't tell me there's no one to look in on me over there."

"But, Mother, you're too old to live alone."

"No, I'm not. I'm too old to live any other way."

"But we don't want to be worrying about you. We want to look after you. Why, Mother, some days you can't even get out of bed."

"Because I don't have to," Laura replied. "Oh, there will be mornings when breakfast in bed will sound like heaven, but, until I'm actually there, it's better for me to get up. Who's feeding my birds, I'd like to know?"

Her other children were invited to dinner that night in order to persuade her to change her mind, to live in the city at least, if not with one of them.

"I can't afford to live in the city," Laura answered them. "And I don't want to."

They were irritated with her, even angry, but finally she could begin to see them giving way, giving up, washing their hands of her. They were dim-witted city people, every last one of them. Nobody on the island would be surprised to see her back. With the insurance money, she could be a paying guest in Lyvia's spare room until the new house was built, a project which would provide much needed work for Jim O'Hea and several others. It was not just Laura's way of saying thank you. It was getting back into the rhythm of give and take, the rhythm of living.

"I'll be going back," Laura told her children, "next week."

They're Mothering Me to Death

BY

EILEEN HERBERT

JORDAN

~

ake two teaspoons," she says, "every four hours." Obediently I open wide and she feeds me a teaspoon of some ruby liquid from a bottle in her hand. "It will control the cough, but it will loosen it, too. Just be sure you buy the right kind. Do you want to take this label with you so you'll know?"

No, I don't. I want something, though. I am overwhelmed by a feeling of déjà vu gone mad, and I want it to go away. You see, the young woman standing there, pointing a spoon at me, issuing directions, defining my cough medicine, is my *daughter*. This is the child whose young chin I once held firmly until she swallowed her vitamins, the one whose whole medical history used to belong only to me.

"I took two aspirin when I woke up," I had told her when she noted my muffled voice. It is a summer cold and I will undoubtedly recover.

She is a stay-at-home mother now and as a result is queen of the TV commercials, a princess of patent medicine. "Well," she says thoughtfully, "you could certainly do better than that." She opens her medicine cabinet and ticks off a body of remedies for me. Neo-Synephrine, Primatene, Dristan, Actifed, Contac, Benadryl, Ascriptin...there are more, but I don't hear them. By this time I am too busy plotting my escape.

Just when you think you have survived the pitfalls of parenting, the very people you nurtured in years past turn around and begin nurturing you. Surviving their efforts to insure your survival is no mean feat.

Medication isn't the only threat you face. Remember, your children belong to the exercise and pure food generation. Food is not something wonderful any-more. It is instead an amalgam of carbohydrates and magnesium, potassium and zinc and fiber. Entire categories are fraught with peril. Take red meat.

It is New Year's Eve. I am sitting with my son and his wife in a restaurant in Old Town in Albuquerque, New Mexico. The room is so authentically Western that I feel like Miss Kitty; when I move I'm sure my skirts will swish. I open the menu and the special of the night leaps out at me: Prime Ribs of Beef au Jus.

"The special is prime ribs," I say after a minute.

They both nod. They look, well, not horrified, just pained.

"I know it's red meat," I say. But I have come a long way to visit them and they are polite. Besides, although they don't know it, I am prepared to brazen it out; after all, I was courted in some of the better steak houses in New York. I order it. I eat it with guilt. I recall that I had ignored their advice before on health matters (bee sting therapy for my backache; a pillow filled with mugwort, white sage, hops, and bear root for insomnia). And here I was, failing food. As I plowed my way through my cho-lesterol hell, I got the distinct impression from the sidelong glances my son was giving my plate that he feels my days are numbered.

I will stand my ground with food the way I did that night, though sometimes I simply eat hamburger or Häagen Dazs in private. When it comes to exer-cise, however, I find it best to lie. You can be in deep, deep trouble if you don't. I tell them I just love the fitness program for older women they have found; other-wise I know they will surely find another that is worse (I note, heaven help me, an advertise-ment for a video that is called *Buns of Steel*). Or they will come up with another machine I should buy. As one who has never owned a car without power steering, why I would want to meddle with these mechanical devices that make you push and pull and strain and hop and trot and rotate, I have no idea.

"The best thing for you would be a Stair-Master," my son says.

I spent years of my life schlepping up and down stairs. I had a laundry room in the base-ment and bedrooms on the third floor. There was a straw basket at the foot of the stairs where family members were supposed to drop their dirty clothes, then carry the clean ones up when they went to their rooms. The only one who ever carried that basket upstairs was me. Now I live in an apartment, I glide from room to room, and I may never climb a stair again.

"It's too bad I don't have room for one," I say to him. "That's the trouble with living in an apartment."

Yes, I will cheat. I will feed their fiber cereal to the birds. I will pour their medicine down the drain. I will lie and connive and conceal my vices. But, do you know what? I don't mind. While my contemporaries are reading the obituary pages and cataloguing their ailments in every conversation, I find hope in the fact that my children are behaving as if, handled right, I could live forever—whether I want to or not.

In the small New England churchyard that contains my family plot, there is a tall obelisk with the names of the dead inscribed. In the mid-nineteenth century, one branch of the family recorded the birth and the death of three sons named Thomas. The first one died when he was two, the second at four years of age, and the third when he was 65. I have always had a certain admiration for that long ago lady who just kept at it. But she never knew, I guess, that it doesn't work that way. With each child, you have only one chance. I had only one with each of my children, and now I'm the only mother they will ever have. So let's go for the run, shall we? Just don't tell me the kind of sneakers I have to buy.

Sandwiched

~

BY

SUZAN

MILBURN

musty bandages
brown and stretched
from thigh to heel
hold her reclined in the Lazy-Boy
ready for the twice a day ritual

I unroll her flesh
where varicose veins have been pulled
like worms
from her body

then I tightly re-wind
she thanks me   again

all that is
or ever was
are these rooms
my mother and me
home together four days now

I tell her I want to care
given for care received
yet feel like a trapped volunteer

in the evenings I drive home
my children bound towards me
lap words on my skin
tongues rough with tenderness

my husband surfaces
amongst debris of the week,
watches me insist through
my mail and messages catches me
for a brief embrace

driving back in the summer dark
I relish the red lights
savour my aloneness
as it spreads from open windows
throughout the car
I am the weight on a pendulum
and feel the pavement
as it speeds beneath the wheels

The Day I Married...

BY

JO

CARSON

The day I married, my mother
had one piece of wedding advice:
"Don't make good potato salad," she told me,
"it's too hard to make
and you'll have to take something
every time you get invited somewhere.
Just cook up beans, people eat them too."

My mother was good at potato salad
and part of the memories of my childhood
have to do with endless batches made
for family get-togethers, church picnics,
Civitan suppers, Democratic party fund raisers,
whatever event called for potato salad.

I'd peel the hard-boiled eggs.
My mother would pack her big red plastic picnic bowl
high with yellow potato salad (she used mustard)
and it would sit proud on endless tables
and come home empty.

What my mother might and could have said
is choose carefully what you get good at
cause you'll spend the rest of your life
doing it. But I didn't hear that.
I was young and anxious to please
and I knew her potato salad secrets.

And the thousand other duties
given to daughters by mothers
and sometimes I envy those women
who get by with pots of beans.

Good
Housekeeping

BY

BAILEY

WHITE

t was the middle of November, just a month before the wedding, when my mother announced that she was going to invite the family of our cousin's bride to Thanksgiving dinner at our house. "They need to get to know us on our own ground," she said. She reared back in her reclining chair. "You girls can help with the cooking. Let's see, there will be ten of us, and six of those Mitchells" (the bride's family).

My mother was sitting in the kitchen, dammed in by stacks of old *Natural History* magazines. Behind her a bowl of giant worms, night crawlers, was suspended from the ceiling. She uses worm castings as an ingredient in her garden compost, and she keeps the worms in the kitchen so she can feed them food scraps.

My sister and I didn't say anything for a while. I watched the worms. Every now and then one of them would come up to the edge of the bowl, loop himseif out, swag down—where he would hang for an instant, his coating of iridescent slime gleaming—and then drop down like an arrow into another bowl on the floor. My mother had an idea that the worms missed the excitement of a life in the wild, and she provided this skydiving opportunity as an antidote for boredom.

My sister was eyeing the jars of fleas on the kitchen counter, part of an ongoing experiment with lethal herbs.

Those worms, or their ancestors, had been there my whole life, but somehow, until this moment, it had not seemed odd to have a bowl of night crawlers getting their thrills in the kitchen.

"Worms," I whispered to myself.

"Fleas," my sister whispered.

My eyes fell on a rusty 1930s Underwood typewriter under the kitchen sink. It had been there as long as I could remember, the G key permanently depressed, the strike arm permanently erect. My sister and I exchanged a look.

"What is that typewriter doing under the sink?" I asked flatly.

"Why on our own ground?" said my sister.

"Let's see, we'll have your Aunt Thelma's sweet potato crunch, and Corrie Lou's cranberry mold," my mother said.

Beside the typewriter was a guide to the vascular flora of the Carolinas, a turtle skull, and a dog brush. There were hairs in the dog brush, black hairs. Our dog Smut had died fifteen years ago. I thought about the bride's family—nice people, we were told, from Bartow County—walking into this house on Thanksgiving Day.

"Welcome to our home," my mother would say. And she would lead them over the stacks of books, through the musty main hall, and into a twilight of clutter. They would clamp their arms to their sides and creep behind her with their tight lips and furtive eyes, past rooms with half-closed doors through which they would glimpse mounds of moldy gourds, drying onions spread on sheets of newspaper, broken pottery in stacks, and, amazingly preserved, my grandfather's ship model collection. From one room a moth-eaten stuffed turkey would blindly leer out at them. "Storage!" my mother would explain cheerfully.

The guests would be settled on the front porch, where they would gaze hollowly down into the garden while our mother explained the life cycle of the solitary wasp who made his home in one of the porch columns. My sister and I would pass around plates of olives and cheese brightly, trying to keep a lilt in our voices and making the guests feel "at home."

"You can't do it!" my sister exploded. "We can never get ready in time!"

"What is there to get ready?" our mother asked innocently. "Just the food, and we'll do that ahead of time. You should always do the food ahead of time, girls," she instructed us. "Then you can enjoy your guests."

"Mama!" my sister wailed. "Just look at this place!" She gestured wildly.

"What's wrong with it?" My mother peered out at the room through a haze of dust. Behind her, another worm dropped.

"Just look!" Louise threw her arms wide. "The clutter, the filth..." She spied the rows of jars on the counter. "...The fleas!"

"Don't worry about the fleas, Louise," our mother reassured her. "I am working on a new concoction, based on myrtle and oil of penny-royal. I may have the fleas under control by Thanksgiving."

Louise sank into a chair and looked our mother in the eye. "Mama," she began, "it's not just the fleas. It's..."

But I had come to my senses.

"Stop, Louise," I said. "Get up. We've got a weekend. We'll start on Saturday."

Louise arrived at dawn, the Saturday before Thanksgiving, loaded down with vacuum cleaners, extra bags and filters, brooms, mops, and buckets.

Mama was sitting in her chair in the kitchen, eating grits and making feeble protestations. "You girls don't have to do this, Bailey. I'll sweep up Wednesday afternoon. Then on Thursday there will just be the cooking."

"I know Mama," I said, "but we want to do a good job. We want to really straighten up. You'll be glad when it's all done. Eat your grits." I didn't want her to see Louise staggering out with the first load for the dump: a box of rotten sheets, some deadly appliances from the early days of electricity, and an old mechanical milking machine with attachments for only three teats.

Mama would not let us throw out a box of old photographs we found under the sofa—"I may remember who those people are some day"—or the lecherous old stuffed turkey with his hunched-up back and his bad-looking feet. "It was one of Ralph's earliest taxidermy efforts," she said, fondly stroking the turkey's bristling feathers down. And she let us haul off boxes of back issues of the *Journal of the American Gourd Growers' Association* only if we promised to leave them stacked neatly beside the dumpster for others to find. But she got suspicious when she caught Louise with the typewriter.

"Where are you going with that typewriter, Louise?" she asked.

"We're going to throw it away, Mama."

"You can't throw it away, Louise. It's a very good typewriter!"

Louise was getting edgy. "Mama, it's frozen up with rust and clogged with dust. None of the moving parts moves. And they don't make ribbons to fit those old typewriters anymore."

"Nonsense," said Mama, "You put that typewriter down, Louise. It just needs a little squirt of oil. Bring me the WD-40."

Louise put the typewriter down with a *clunk*. I brought a can of WD-40, with the little red straw to aim the spray. Mama put on her glasses, pursed her lips, and peered into the typewriter. *Skeet! Skeet!* She went to work with the WD-40 and a tiny, filthy rag. "You girls are throwing away too much," she said.

By midafternoon we began to feel that we were making progress. We could see out the windows, and we had several rooms actually in order. We had found our brother's long-lost snakeskin collection and the shoes our great-aunt Bertie had worn at her wedding; a dusty aquarium containing the skeletons of two fish; and under a tangle of dried rooster-spur peppers and old sneakers, a rat trap with an exquisitely preserved rat skeleton, the tiny bright-white neck bones delicately pinched. "Just like Pompeii," Mama marveled.

By the end of the day we had cleared the house out. What had not been thrown away was in its place. I had dropped a drawer on my foot, and Louise was in a bad mood. Mama's glasses were misted with WD-40. We sat down in the kitchen and drank tea.

"What I want to know is, where are the priceless heirlooms?" asked Louise. "You read about people cleaning out their attics and finding 200-year-old quilts in perfect condition, old coins, cute kitchen appliances from the turn of the century, Victorian floral scenes made of the hair of loved ones. What kind of family are we? All we find is bones of dead animals and dried-up plants. Where are the Civil War memorabilia, the lost jewels, the silk wedding dresses neatly packed away in linen sheets and lavender?"

"Well," said Mama, "you found your brother's snakeskins. And I think this rat skeleton is fascinating. How long must it have been there?"

"Don't ask," moaned Louise. "I'm going home."

On Sunday we dusted everything, swept, vacuumed and mopped the floors, washed the windows, and laundered the curtains, rugs, and slipcovers. By nightfall the house was ready.

"You girls have certainly struck a blow," Mama congratulated us. "This place is as clean as a morgue." We left her sitting in her chair with the worms, the typewriter, and the last three surviving fleas.

I walked out with Louise. "She looks a little forlorn," I said.

"She'll get used to it," Louise declared. "And the Mitchells will never dream that we are peculiar!"

Thanksgiving morning. Louise and I divided up the cooking. She made the sweet potato soufflé and the squash casserole, and I cooked the turkey and made bread. Mama spent the morning in her garden picking every last English pea, even the tiniest baby ones, because we knew we would have our first freeze that night.

At ten o'clock we set the table. For a centerpiece Mama put some pink and white sasanquas to float in a crystal bowl, and the low autumn light came slanting in through the windows onto the flowers and the bright water. We had built a fire in the stove, and the heat baked out the hay-field fragrance of the bunches of artemisia hung to dry against the walls. The floors gleamed. The polished silverware shone. Beneath the sweet fall smells of baking bread and sasanquas and drying herbs I could just detect the faintest whiff of Murphy Oil Soap. Louise and I stood in the middle of the living room and gazed.

"The furniture looks startled," Louise said.

"It's beautiful," I said. "And here they are."

"Welcome to our home. We're so glad you could come," Mama was saying to the Mitchells. "Come out onto the porch. You will be interested to see the wasp who lives there. It's a solitary wasp, quite rare...I know it looks a bit cleared out in here; my girls have been cleaning. Bailey, Louise, come and meet these Mitchells."

We sat on the porch for a while, bundled in coats, and watched the last petals of the sasanquas drift to the ground. Mr. Mitchell examined the neat, round hole of the solitary wasp with some interest.

"Do you have a knowledge of the hymenoptera, Mr. Mitchell?" Mama asked. And she was off.

Mrs. Mitchell had smiley eyes and a knowing look. She leaned over to Louise and me. "It's the cleanest house I've ever seen," she whispered. We were friends. Louise and I took her to the kitchen to help with the food.

Other guests arrived—our brother and his family, aunts and uncles, and the bridal couple. The house was full of talk and laughter. We brought out food and more food. Everyone sat down.

Then, "Where's Daddy?" asked the bride.

Sure enough, an empty chair...two empty chairs.

"Where's Mama?" asked Louise.

"On the porch?"

No.

"In the kitchen?"

No.

"Everyone please start. The food will get cold," I said. "I'll go find them."

Outside, the temperature was dropping. This was the last day the garden would be green. I wandered along the path, following the scent of bruised basil until I heard voices way in the back of the yard.

Mr. Mitchell: "...and this is?"

"*Franklinia altamaha*, Mr. Mitchell, and quite a spectacular specimen, if I do say so."

"The famous Lost Franklinia of John Bartram," Mr. Mitchell murmured reverently, gazing up into its branches. "I have never seen one."

The sun shining through the crimson leaves of the Franklinia lit up the air with a rosey glow. Mr. Mitchell was holding her arm in his and gesturing with her walking stick. She was cradling some stalks of red erythrina berries in their black pods.

Mr. Mitchell turned slowly and looked over the garden. "Silver bell, shadbush, euonymous, bloodroot, trillium"—he named them off. "Mrs. White, I've been collecting rare plants and heirloom seeds all my life, and I've never seen anything to equal this."

"It's an old lady's pleasure, Mr. Mitchell," said my mother. "Now wait till I take you to the dump and show you my bones. Louise threw them out," she whispered hoarsely, "but I know right where they are. We'll get them tomorrow, if you're interested. You will be kind enough not to mention it to my girls."

"It would be my extreme pleasure to see your collection of bones, Mrs. White," said Mr. Mitchell. And slowly he led her out of the pink glow and back to the party.

The next day Louise came over, and we went to sit in the kitchen and drink hot chocolate with Mama and congratulate ourselves on a job beautifully done. But Mama was not in her chair. There was a note on the kitchen table. It was typewritten. Every letter was clear and black and even.

> Sorry I missed you girls. Mr. Mitchell and I have gone on a little errand. Make yourselves some hot chocolate.
>
> Love,
> Mama

Daddytrack

BY

JOHN BYRNE BARRY

It is not news that fathering has changed over the past generation or two, that today's father is more involved in child care than his father was. But although the involved and nurturing father is becoming more visible and acceptable, he is still generally regarded as a helper in the world of child care and housekeeping. He pitches in. He helps *when he wants to.* Fathers are volunteer providers; mothers are the staff. The household with two "staff parents" is still rare, especially when it requires father to cut back on work.

TV commercials show fathers in business suits ducking out of conferences to attend their children's school plays, but we do not see them ducking out of work for the plain-old-vanilla caretaking of their children—the "quantity time" stuff. Support for fathers' involvement in the day-to-day labor of raising children is growing in the work world with the speed of a glacier. Whereas employers may not allow a mother to work part time, they do understand her reason for wanting to. A man who wants to work part time in order to care for his children is looked on with suspicion or, at best, with amusement.

Laurie and I decided to share child care before we were married, well before our son Sean was born. Neither of us wanted to stop working, nor did we want to miss out on this exciting period of newborn growth by being the full-time bringer-home-of-bacon.

In the months before our son was born, we both went to our employers and negotiated a reduction in working hours. We were able to juggle our time so one of us could always be the primary caregiver. Our employers were supportive, allowing us the flexibility to work at home and not balking at our patchwork schedules. After Sean was born, I worked long hours two days a week while Laurie stayed home with him. On Tuesdays, I was home on the range. On Wednesdays and Fridays we each worked half a day and spent the other half with Sean.

We managed this arrangement for 17 months and then suffered a temporary setback. A restructuring at my workplace and a promotion I never asked for sent me across the bay to San Francisco...full time. Despite an impassioned speech about why I should maintain a flexible schedule, I found myself being a commuter dad, away from home 11 hours a day, five days a week.

The changes in my work schedule provoked changes in Sean. He began to cling more to his mother. And on the one evening a week that I soloed, he would wander through the house baaing "Mama" like a lost sheep and stay awake until she returned home around 10.

After four months of full-time work—and a lot of grumbling—I found a flexible, 80-percent-time job. Although our situation is tighter and less flexible than it was originally, we have recovered our sense of balance. My relationship with Sean has improved dramatically.

We've been lucky, yes, but the determining factor in our success has been *asking for* and *looking for* suitable work arrangements. Although part-time professional work, job sharing, and working-at-home situations are still scarce, they have nevertheless entered the vocabulary. According to proponents of alternative work arrangements, the most successful route to a flexible situation is to create it yourself and our experiences attest to that. For now, this is more easily accomplished by workers who have already established the trust and respect of their employers.

Our new arrangement is working well for the entire family. Sean is a good-natured, energetic, outgoing, secure child who laughs more than he cries. Laurie loves the balance of playing with him in the mornings and then working in the afternoons. I also enjoy balancing the intellectual demands of work with the emotional demands of parenting. And the quantity of time we have together frees Sean and I from always having an agenda: We hang out, take aimless walks, play on the swings, eat ice cream, or sit on the corner and watch cars and trucks go by.

Although Laurie and I continue to strive toward equality in

our parenting, we realize that it has never been *truly* equal—partly because Laurie logs more hours with Sean, but mostly because of the strong biological attachment between mother and child. Among couples who are committed to true equality in child care, mothers sometimes opt to nurse for just a few months or not at all. For us, the importance of breastfeeding took precedence over any ideological commitment. In the early months, when Sean nursed a lot, our division of duties was weighted toward me doing most of the household stuff and Laurie doing most of the baby stuff. As Sean grew older and began to nurse less, we adjusted our division of labor. We seem to be mirroring what physician Kyle Pruett calls the Jack Sprat theory of parenting: Our contributions are not similar, but rather complementary.

One way we keep current is by having "business meetings" on Friday evenings after Sean has gone to bed. This gives us an opportunity to check how we are doing with our responsibilities as well as to synchronize calendars, schedule child care, balance the checking accounts, and so on.

All is not milk and honey, of course. Quantity time with Sean translates into a shortage of time for his parents—for romance, for play, for sleep, and for keeping house. We've also made sacrifices—we're renting a house instead of buying one. Career advancement is on idle. And although I know in my heart that nurturing a strong family is more important than owning a house or having a fat paycheck, it's tough to ignore all the people cruising by me on the fast track.

Actually, the career pressure has been of less concern than the lack of peers making similar choices. One of the questioning voices in my head has nagged, "If it's so right, how come more fathers aren't doing it?" I have often felt isolated as a father, a daddy lost in mommyland, especially on weekdays, when I find myself in a gym or on a playground filled with babies and their mothers.

When I was a child, no boy ever said he wanted to grow up to be a father. Perhaps when Sean and other boys of his generation start thinking about what they want to be, some will decide to be fathers. And perhaps, when they ask their employers for a reduced schedule in order to care for their children, they will be met with a knowing smile.

His Kitchen

BY

ANNHARTE

My father was my mother. He took over
cooking and childcare when she left.
At first, our food came from a can.
He wouldn't let me near the kitchen.
I had to learn to cook at school.
He improved. I asked friends over.
He didn't mind. He heaped up potatoes
and gave us canned fruit for dessert.
Only for a short time, did we go out
almost every night to a restaurant.
Even now, I know I am in his kitchen.
A paint scraper sits with the utensils.
I want to put it back with the tools.
It is his egglifter so I know better.
Holiday dinners he cooks and I make gravy.
Hard to forget he's both mother and father.

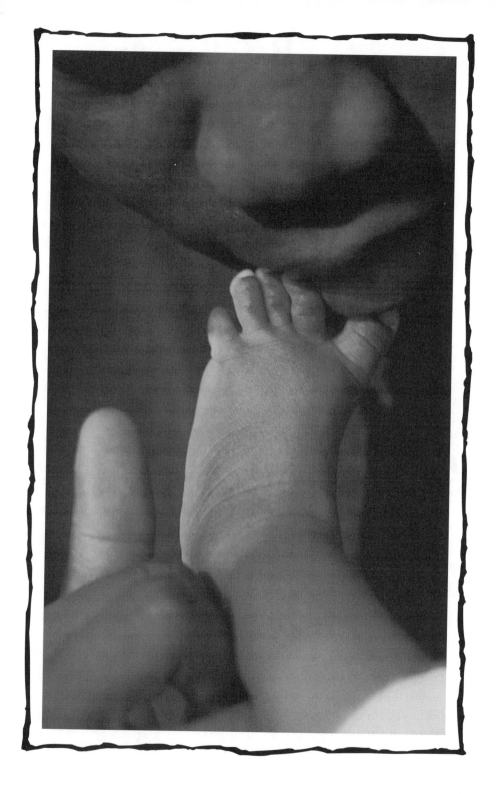

How Poverty Affects My Kids

~

BY

ANONYMOUS

I guess for me it is a little easier because I am married. I have my husband beside me. He had a fair paying job with the city until the cutbacks started. When he was laid off, we were waiting for the arrival of our third child. We were devastated. We hoped that before the baby was born, he would be back to work. Boy, were we wrong. We now have three children, two girls ages sixteen and three years, and one son age eight. It has been three years since my husband has had any steady employment. He has had odd jobs, but nothing ever seems to be permanent.

I think the first year was the hardest to deal with on welfare. We had to get used to living on a fixed income of $798 per month. Thank God for B.C. Housing [government-subsidized housing available to a small number of low-income people]. Our rent is at least reasonable—$217 a month. After we pay our utilities there never seems to be enough left over for food or clothing. I am becoming very adept at shuffling money. The old adage "Steal from Peter to pay Paul" is true. It's tough. It's really tough. Once we get our cheque at the end of the month, we have food in the house for about two weeks and then it's gone. So is the money. So many nights I put the kids to bed and they are saying, "But Mom, I'm still hungry." It just breaks my heart to watch them go to the fridge and leave it empty-handed. Or to grab a carrot and run it under water to see if it will stiffen up. By the end of the month there is usually only carrots and onions left in the fridge; all the good stuff has long been eaten. The milk lasts about two to two and a half weeks, and then that's it for the rest of the month. Same with the eggs and cheese. The fruit usually lasts one week. Canned goods last a little longer, and thank God for the food bank; they make things go a little bit farther. The powdered milk helps, for baking and things. My kids just can't seem to make themselves drink it, so I usually only use it for cooking. I brought them up on good milk and no matter how hard I try to make them drink the powdered stuff, they still choke and cry and carry on. They will take it in their cereal if I can mask it with a little extra sugar or cinnamon. My son has gone to school many a morning crying because I made him eat the cereal with "that milk" on it. I'm sorry, honey, I try.

We try to budget the money, but no matter how hard we try there just doesn't seem to be enough to feed my growing family, not nutritionally anyhow. They are good kids. They hardly ever complain, but I can see it in their faces when I give them "instant noodles again!"

My eldest daughter turned sixteen in December. Also in December, she decided to quit school. She has been a troubled child for a few years now. One of her reasons for quitting was clothing. She felt very uncomfortable going to school in "those holey jeans." Let's face it, even when I went to school there was that competition on who's wearing what. It's still there, even though now it is blue jeans and T-shirts. Those jeans are expensive, as are the runners they wear. She couldn't compete and felt that quitting was easier. She never complained to me, she just quit. Period. Some Sweet Sixteen.

Last month she came to me and told me she was pregnant. She is going to keep the baby, and we are going to stand behind her. We will try and help her out the best we can. She has lost her self-esteem and I am going to try with all my might to help her re-find herself. I just pray to God that my husband finds a steady job before the baby is born. I don't blame welfare for my daughter's plight, but I feel that if things were just a little bit easier for us, maybe she would feel better about herself and things would have worked out differently.

I worry about my eight-year-old son. We live in a housing complex. There is an awful lot of drug use in our area. The kids he hangs out with are good kids, but they are all so easily led. I wish I had the extra money to get him more involved in sports; he played soccer for one season, but we couldn't afford another season. I pray to God that things will change in our lives before it's too late for him.

Our youngest daughter's biggest thrill of the week is Food Bank Wednesday. She goes with me and she usually gets free candy from them. I don't want her to get as frustrated as my other kids are. I hope things will change before she gets too much older.

I worry about the kids' teeth. We are not eligible for free medical or dental coverage. The medical is okay because we can purchase it privately, but not the dental. The kids haven't been to a dentist in over three years. My eldest daughter has had a cavity in her tooth for the last three years, my son needs dental work, and the youngest has never even seen a dentist. I had their names on a list at the university dental clinic, but when it came time for them to go (on two separate days), we just didn't have the money to get them there. My welfare worker told me of a place where we can get work done for a discounted price. The discount is 20% off, but on welfare there is no room for extras, even if it was 90% off. I pray that they can just hang on a little bit longer.

How can I define the feeling I have every night when I go to bed? Some nights I just can't wait to go to bed. That's when I can finally take off that "happy face" and feel the way I really feel. No more putting on happy fronts for the kids or the neighbours, when all you really want to do is cry. Nighttimes are the worst for me. That's when all my thinking and worrying is done: "How will we get through another day? What will I feed them tomorrow? Who can I borrow from to make it just one more day? When will it end?" Usually through the sheer frustration of worry and crying I fall asleep, only to wake up to another day.

People never really think of what it's like to be poor until they are poor themselves. It's a sad fact but it's true. It doesn't help for some-one to try it out for a day, or a month, or even six months. They have to live it. My husband is not one of those "welfare bums." He tries, he tries really hard. He doesn't like depending on the government to sup-port us. If he could find a job today he would take it. But that is a pret-ty big "if" in this day and age.

There must be some way out of this nightmare. It's tough. It's really tough. We keep praying that tomorrow, maybe tomorrow, things will change, something will happen to make it a little easier until that job comes through. Someday…

The Kids Are All Right

BY

SUSAN

FALUDI

~

What should be most curious to anyone reviewing the voluminous research literature on day care is the chasm between what the studies find and what people choose to believe. Despite a pervasive sense that day care is at best risky for children and at worst permanently damaging, much of the research indicates that if day care has any long-term effect on children at all, it has made them somewhat more social, experimental, self-assured, cooperative, creative. At the University of California at Irvine, Alison Clarke-Stewart, professor of social ecology, found that the social and intellectual development of children in day care was six to nine months ahead of that of children who stayed home.

The research on day care points to other bonuses, too. Day-care kids tend to have a more progressive view of sex roles. Canadian researchers Delores Gold and David Andres found that the girls they interviewed in day care believed that house-work and child care should be evenly divided; the girls raised at home still believed these tasks are women's work. We are reminded constantly by the press that children in day care turn out to be too "aggressive." But researchers point out that what is being billed as aggression could as easily be labeled assertiveness, not at all a bad quality in a child.

As for the supposed and much-publicized day care child abuse "epidemic," a three-year, $200 000 study by the University of New Hampshire's Family Research Laboratory found in 1988 that if there is an "epidemic" of child abuse, it's in the home—where children are almost twice as likely to be molested as in day care. And, ironically, the researchers found that children were *least* likely to be sexually abused in day care centers located in high-crime, low-income neighborhoods (there tends to be more supervision in these centers). Despite frightening stories in the media, the researchers concluded that there is no indication of some special high risk to children in day care.

Although many of the celebrated tales of day-care workers molesting children have turned out to be tall tales, we continue to believe their message. "The consequences of all the negative play in the press about the McMartin case were really quite dramatic," says Abby Cohen, managing attorney of the Child Care Law Center, referring to the 1984 sex-abuse scandal at the McMartin Pre-School in Manhattan Beach, California. "Because, unfortunately, up until then there hadn't been very much play about child care, period. So it was terribly detrimental that the first real wide attention to it was in such a negative light."

For children of poverty, day care may be their ticket out of the ghetto. The studies find that the futures of low-income kids brighten immeasurably after a couple of years in day care. The Perry Pre-School Project of Ypsilanti, Michigan, followed 123 poor Black children for 20 years. The children who spent one to two years in preschool day care, the researchers found, stayed in school longer, were not as prone to teenage pregnancy and crime, and improved their earning prospects significantly. A New York University study of 750 Harlem children came up with similar results: The children enrolled in preschool were far more likely to get jobs and pursue education beyond high school.

Presented with the evidence, some day-care critics will concede that preschool may have a negligible negative effect on toddlers, but then they move quickly to the matter of infants. So 3-year-olds may survive day care, they say, but newborns will suffer permanent damage. Their evidence comes from two sources. The first is a collection of studies conducted in the 1940s, '50s, and '60s in France, England, and West Germany. These studies concluded that infants who were taken from their mothers had tendencies later toward juvenile delinquency and mental illness. But there's a slight problem in relying on these findings: The studies were all looking at infants in orphanages and hospital institutions, not day-care centers.

The other source frequently quoted is the much-celebrated turnabout by Pennsylvania State University psychologist Jay Belsky, once a leading supporter of day care. In 1982 Belsky had reviewed the child-development literature and concluded that there were few if any significant differences between children raised at home and those in day care. Then in September 1986 he announced that he had changed his mind: Children whose mothers work more than 20 hours a week in their first year, he said, are at "risk" for developing an "insecure" attachment to their mothers. Belsky's pronouncement provided grist for the anti-day care mill—and was widely reported. What did not receive as wide an airing, however, is the evidence Belsky cited to support his change of heart. Two of the studies he used flatly contradict each other: In one of them, the study's panel of judges found the infants in day care to be more insecure; in the other, the panel found just the opposite. The difference in results was traced to the judges' own bias against day care. In the study where the judges were not told ahead of time which babies were in day care and which

were raised at home, the judges said the children's behavior was indistinguishable. In the study where they did know ahead of time which babies were in day care, they concluded that the day-care children were more insecure.

It is this bias that makes our day-care terrors so intractable. If the feeling comes from the gut, if it is an internalized, strictly personal belief, then its truth must be of a higher order, unassailable by any number of studies. But what many people fail to see is how our seemingly personal perceptions of day care are not so personal after all: how they have been shaped by forces that have little to do with gut instinct. Our opinions have been hammered by years of relentless anti-day care and anti-working-mother rhetoric from the Reagan and Bush administrations and from the media, where bashing day care seems to be a sanctioned sport. (A few headlines from well-read magazines: "Mommy, Don't Leave Me Here!: The Day Care Parents Don't See." "When Child Care Becomes Child Molesting: It Happens More Often Than Parents Like to Think." "Creeping Child Care... Creepy.")

Other cultural forces are at work here, too. We suffer a

compulsion to replicate our childhoods, no matter how unpleasant those early years might have been. If our mothers stayed home, that must be the "healthy" way. What we forget is that it's only been since the 1940s that public opinion has so insistently endorsed the 24-hour-mom concept. The Victorians may have kept their women at home, but not for the sake of the children. "An educated woman," writer Emily Davies advised mothers in the 1870s, "blessed with good servants, as good mistresses generally are, finds an hour a day amply sufficient for her domestic duties." An early version of quality time.

The paranoia may ease once the younger generation reaches adulthood; they are not freighted with the same cultural assumptions about child care as those weaned in the 1950s. When I brought up the matter of child rearing to teenagers at Lowell High School in San Francisco, they all seemed to favor day care more than their parents do. How come? "Well," one 17-year-old girl reasoned, in what turned out to be a typical response, "I went to day care when I was little and I had a really good time."

Kids Have Rights Too!

BY

MARCIA

KAYE

The mention of children's rights is enough to unnerve even the most loving parent. "Children's rights?" we mutter, "What about parents' rights?" We've all become very conscious lately of our legal rights as workers, voters, spouses and citizens. Our Constitution has enshrined these rights for adults. But rights for children is another story. Many of us harbor a fear that enshrining rights for children will somehow weaken parental authority and undermine the entire structure of the family.

On the contrary, the aim of children's rights is to strengthen families and communities, as stated by the United Nations Convention on the Rights of the Child. This international legal agreement itemizes basic rights for the world's children. These include the right to a name, a nationality, health care, education; the right of protection from abuse; the right to practise their own religion, language and culture; and the right to a safe, secure family life.

While the idea of children's rights may seem radical, it's hardly new. It was first discussed by the original League of Nations back in 1924. Canada, which played a leading role in getting the convention off the ground, ratified the agreement in 1991. More than 170 countries have now signed it.

"The convention is a very important document because, for the first time, it sets out a conceptual framework that says children are human beings, too, not products owned by society," says Rix Rogers, director emeritus of the Institute for the Prevention of Child Abuse.

While some people believe that the convention came about because society values children more than ever, Toronto psychiatrist Dr. Paul Steinhauer says that in Canada, children are not highly valued. A century ago, children had an economic worth: they could be counted on to help run the family farm or take care of aging parents. Today's children, who often remain financially dependent on parents until at least their mid-20s, are viewed as liabilities. "We really don't take children seriously," says Steinhauer, chair of the steering committee of the Coalition for Children, Families and Communities. "We assume that all we have to do is feel benignly toward our own kids and everything will be fine."

One of the most important concepts in the UN convention is the notion of community responsibility toward children. The old-fashioned theory that children are solely the responsibility of their individual family no longer works in today's society. One in nine children in Canada depends on the charity of others just to eat. More than three-quarters of families need two incomes to stay out of poverty, leaving child care to people outside the family. Despite all our labor-saving devices, parents today have less time for family activities than they did a generation ago, causing what sociologists call a "family-time famine." Although we still cling to the traditional concept of family—children living with two parents, one working and one

staying at home—in 1991 only 12 per cent of Canadian children lived this way.

"We need a lot more than simply saying to parents, 'Why don't you shape up and take care of your children?'" says Rogers. Faced with the tremendous stresses, financial and otherwise, of modern life, families need the support of knowing that governments, businesses, schools and communities join with them in putting children first.

This means that society must support initiatives such as prenatal counselling, quality day care and parenting taught in schools. Ensuring that children get off to a decent start in life, with their basic physical and emotional needs met, is less costly and far more beneficial than trying to intervene later, when damaged children become substance abusers or young offenders. "Trying to turn a kid around after the damage has been done is like trying to turn around a 14-wheel tractor-trailer that's going 100 kilometres an hour," Steinhauer says.

Another striking feature of the UN convention is that it gives children the right to express their views freely in decisions that affect their well-being. "We always say we're doing things in the best interests of the children, yet we're so adult-centred that we never think to ask the children," says Ilze Kalnins, associate professor with the department of behavioral science in the faculty of medicine at the University of Toronto. What children want can be very different from what adults think they want. When the Canadian Mental Health Association wanted to set policy goals for youth, it first consulted a group of adults. But when later consulted, the young people themselves had very different priorities. Security topped the adults' list but was last on the youths' list. Number 1 for youth was to be treated with respect, an item that didn't even make it on the adults' list. "It hammered home in a very dramatic way that when we make decisions concerning youth, we should include youth," says director of programs Bonnie Pape.

This idea of giving children a voice is popping up here and there: a committee of eight- to 13-year-olds helps choose exhibits for a children's museum in Winnipeg; a judge puts a twist on joint custody in Nictaux, N.S., by allowing the children to stay in the house and ordering the parents to alternate living there; children are increasingly consulted by designers of playgrounds, libraries and toys.

Where does all this leave parents? In as vital a role as ever. "This isn't about letting your children slouch on the couch all day with their feet up," Kalnins says stressing that supporting children's rights does not mean supporting permissive child rearing. Far from undercutting parental authority, the UN convention recognizes that parents have the primary responsibility in raising their children and deserve all the help society can offer. "It's a misconception that the convention is pitting children's rights against adult rights," says Sydney Woollcombe, Canadian program manager of Save the Children-Canada. "The whole idea is to help children become more responsible members of society."

Parents' influence is crucial. A recent study shows that young people who experience difficulties with their parents are more likely to engage in health-risk behavior such as substance abuse or irresponsible sexual activity. Another study of 17-year-olds found that parents had a huge amount of influence on young people's decision about their educational future.

Ultimately, the enshrining of children's rights will continue to strengthen the role of parents. For by cherishing children, we automatically increase the value of those who raise them.

The Child Who Walks Backwards

BY

LORNA

CROZIER

My next-door neighbour tells me
her child runs into things.
Cupboard corners and doorknobs
have pounded their shapes
into his face. She says
he is bothered by dreams,
rises in sleep from his bed
to steal through the halls
and plummet like a wounded bird
down the flight of stairs.

This child who climbed my maple
with the sureness of a cat,
trips in his room, cracks
his skull on the bedpost,
smacks his cheeks on the floor.
When I ask about the burns
on the back of his knee,
his mother tells me
he walks backwards
into fireplace grates
or sits and stares at flames
while sparks burn stars in his skin.

Other children write their names
on the casts that hold
his small bones.
his mother tells me
he runs into things,
walks backwards,
breaks his leg
while she lies
sleeping.

The
Good
Girls

~

BY

FRAN

ARRICK

ary Louise opened her bedroom door a crack and peeked through to the living room. There he was, on the couch, looking as though he were asleep, but she knew he wasn't. He was only completely passed out when his mouth hung open and you could hear his raspy breathing clear through to the back of the house. She couldn't hear it yet, so she stayed in her room, not daring to cross to the front door until she was sure he was sleeping soundly.

She tapped her foot impatiently. He was almost always out by late afternoon, leaving her free to her comings and goings. And when he wasn't on the couch, he could be found in The Roadhouse at the edge of town, right on the border between Armenia and Braverlee. But here he was, still partly awake and right in her way.

It was early spring and beginning to be hot, but Mary Louise still wore her dancer's gear of tights and protective legwarmers and had begun to sweat through them. She knew she'd be soaked through before she even got to the Center to start work. She rubbed the end of her ash-blond pigtail across her lips, stared hard at the couch, and willed her father to slip into oblivion.

Her heart sank as he sat up.

"Hey!" he called. His eyes were still closed and she prayed that he was talking in his sleep.

But "Hey!" he said again, and rolled over, facing her door. His knee knocked an ashtray from the pine coffee table. *"Frances?"* he called.

Mary Louise's mother. Frances. She'd run off when Mary Louise was eleven. She'd been gone five years.

"Hey, Frances!" her father yelled. "Where you at?" He lurched into a sitting position and bellowed again. "Get yourself over here, I said!"

Mary Louise decided to chance running for the door. Even if he hadn't passed out yet, he looked too sodden to stop her if she was quick. And by the time she got back, he'd be dead to the world.

Her father sank back down. He was breathing heavily, but only from the effort of having sat up.

I won't be late, Mary Louise said to herself. I just won't be late for this class. Miss Dorothy had fought for her to have this job. Miss Dorothy knew that, at only sixteen, Mary Louise was the best instructor for the little ones the Center could have. She'd only taught three weeks, but it was going so well and she loved it so much!

I'm leaving, Mary Louise said to herself. I'm just going to go.

She was wearing her soft ballet slippers. She wouldn't make a sound crossing the room, if only he'd just keep his eyes closed. But the door—the sound of that door opening and closing—

She took a breath and held it, opened her door wider, and slipped through, taking one tiny trial step into the bigger room. Her father lay with one arm thrown over his forehead, not moving.

Watching him as a cat watches when it's stalking, she propelled her dancer's body across the floor. Only thirteen steps, but it felt like miles, and trickles of perspiration rolled down both her sides.

Her fingers touched the doorknob. Oh, please, she prayed softly. Please.

"Where you goin', Mary Louise?"

She froze.

"I said, where you think you goin'?"

"To work, Daddy." Her voice came out in a whisper and he didn't seem to hear her.

"I ast you where you think you're goin', girl!" he roared at the ceiling.

Mary Louise stood with her hand on the doorknob and her foot on the sill. She cleared her throat. "To work, Daddy," she repeated. "I'm going to work now."

"Not till I say so, you ain't!"

"Go to sleep now, Daddy," she said. "I'll be—"

"You tellin' me what to do? It's the other way round, girl!"

Mary Louise managed a soothing voice. "Now, don't you worry, Daddy. I'll be back to fix your supper, 'bout an hour or two." Eyes on him till she slipped through the door onto the porch. She didn't think he'd moved, thank the Lord. He'd forgotten her already. Still, she closed the door as quietly as she could behind her before scurrying down the wooden steps and through the back lot—the shortcut to the Center downtown.

Downtown Armenia, Georgia, was eight stores including a Fast-Way supermarket, a Baptist and a Methodist church diagonally across the road from each other, and a new wood structure near a green, built by both churches to be used as a teen center, a dancing school, a meetinghouse, a lecture hall, and anything else someone might want to rent it out for.

Miss Dorothy Teaman had one of its rooms three days a week—Monday, Wednesday, and Friday—for her École Pour la Dance, and it had turned out to be a big success. Girls from five to fourteen applied in droves, coming all the way from Brittany, Spencer, and Ravenswood, because it was the only school of its kind and because Miss Dorothy's advertisement proclaimed she had actually danced with the famous Rockettes on the stage of the Radio City Music Hall. So the girls dreamed that if they were really good, maybe some of them could someday kick their legs on that big stage and be as far as they could get from Armenia or Brittany or Spencer or Ravenswood.

Mary Louise was one of the dreamers. With money she'd saved from baby-sitting and cleaning houses, she'd joined the Wednesday afternoon jazz class and had progressed so rapidly that Miss Dorothy had managed to convince suspicious parents that Mary Louise Wattles was not too young or too inexperienced to start their own little ones on the road to fame and fortune. She taught Friday afternoons and got to take two other classes from Miss Dorothy herself, free.

Drenched and panting, Mary Louise raced down the long inner hall that led to the big room at the back, where Miss Dorothy had put up a huge mirror on one wall and a barre along the opposite one. Mary Louise could hear the music. They had begun without her.

Eleven five- and six-year-olds were grouped in rows and bent over at the waist, dangling their arms loosely in front of them. Miss Dorothy, her back to the mirror, looked up.

"Here you are, Mary Louise," she said, and smiled.

"I'm so sorry I'm late."

"No problem, we've just begun. You're awfully out of breath. Want to sit and rest for a minute?"

"Oh, no, ma'am, I'm ready," Mary Louise insisted. "And I won't be late again."

"Now, don't worry about it." Miss Dorothy gave her a long look. "Are you all right, Mary Louise?"

"Oh, yes, I'm fine."

"Well, good, then. They're all yours. And I've brought your record in too. It's right next to the machine. I'll be in the office." She smiled again and left.

Mary Louise stepped to the front, filled now with confidence and relief.

"Listen up, now, girls," she said cheerfully. "I want to see those flat backs now. Everybody. Come on now, Lizbeth. You, too, Patty-Warren. One, two, three, four, five, six, seven, *eight!* All right, now, hit the floor—on your backs—" She did it with them, and the sweat she felt now was a good one, a clean one, one she didn't mind at all.

"Feet flexed in the air," she called, "Now, sit up and touch your toes eight times...five, six, seven, *eight!* Ro-o-ll-ll down, chest in, lie down on count of eight. Patty-Warren, why you sittin' up, honey?" She looked over at a small girl, trying to hide behind a bigger one in the front row.

"Patty-Warren?" Mary Louise repeated. The child was new in town, the only daughter of an architect and his wife, here from Atlanta for the planning of a modern shopping mall. It was the talk of the town.

"She don't have to move," a little girl called out. "She'll just git someone to carry her!"

There was some giggling.

Plain mean, Mary Louise thought. They just repeat what the bigger folks say out of jealousy. "Stop that, Sally," she said to the girl who'd called out. "And you other girls, is that any way to treat someone new? Y'all know better. Now, on your feet and let's do this one. Plié...heels up...down. Plié...heels up...down. Ready?"

Out of the corner of her eye she saw Patty-Warren blink her eyes tightly with each count. She was working, but something was wrong.

"You okay, Patty-Warren?" she asked.

"She always slows us up," another girl complained. "Why we have to slow ourselves just 'cause Patty-Warren can't keep up?"

"Now, Marcia, she does keep up. She must be sick today or something, isn't that right, Patty-Warren? Now, let's do those pliés again. Ready?"

Maybe she's just not coordinated too well, Mary Louise thought as she watched the little ones plié, stretch, arch, and bend. There was always one who stood out that way. In her own class it was Dale McMellin, and all the girls called her Dale the Whale, Destined to Fail. Once Dale McMellin had cried, right there in front of everyone, which made it all worse, but little Patty-Warren, ten years younger than Dale and Mary Louise, wasn't crying at all. Her thin little face was pinched in determination and she looked as if she hadn't even heard the others discussing her.

When the warm-up exercises were finished, Mary Louise put on the jazz record that Miss Dorothy had let her pick out, and the girls began the little routine Mary Louise had choreographed for them. This was the part they all loved, because it made them feel like the teenagers they watched in television commercials or at their big sisters' and brothers' parties on Saturday nights. When Miss Dorothy looked in, they were gyrating wildly and grinning from ear to ear. Even little Patty-Warren had closed her eyes and was smiling to herself.

At the end of the class the little girls filed out into the hall, where their mothers waited. Their cheeks were red and their hair damp and curling on their foreheads.

"They looked good doing that routine, Mary Louise," Miss Dorothy said. "You know, I thought at first their parents might object to their doing that sophisticated stuff, but they seem pretty proud of them."

"Yes, ma'am! Marcia Willis said her mama wakes her up sometimes to show off for company!"

"Well, I certainly did the right thing by getting you to teach them. You have real talent, Mary Louise."

The next class, a group of eight-year-olds, had begun to move into the room, some whirling with each other in their own version of *Swan Lake*.

"Come on, let's go into the office for a minute," Miss Dorothy said, taking Mary Louise's arm. "Girls, form a circle with Alice in the middle. And, Alice, you start the warm-ups with the neck exercises—all right, dear?"

She steered Mary Louise into the little room that was supposed to be the church's office but which Miss Dorothy had appropriated for herself, decorating it with her own green leather couch and pictures of herself and all the Rockettes kicking in various costumes and routines.

"Sit down, Mary Louise," she said, pushing some papers off the green leather couch and onto the floor.

Mary Louise sat, but she was nervous, so she began to rub the tip of her pigtail across her lips.

"Honey...I don't want to pry. But I want to help if can."

"I don't need help, Miss Dorothy. Besides, you'll never believe how much you've helped me already, with—"

"I know you're a dancer. I wish you'd been able to start at the age your class is now. But you're awfully good and you'll get better. There are all kinds of things you'll be able to do with it. But no, that's not what I mean. This is a small town—"

"Don't I know it!"

"Well, yes. But— Now look, I don't want to embarrass you. But you hear things, Mary Louise. I know your mother has been gone awhile now, and your father's had a hard time, hasn't he?"

He had a hard time long before she left, Mary Louise thought, but all she said was, "Well, we manage all right, Miss Dorothy."

"All I want, Mary Louise, is for you to know that you have someone to help you if you ever need it. That's all, Mary Louise. Just for you to know that. Do you understand?"

She was putting tap water in the kettle to boil for instant coffee when the blow came, knocking the kettle out of her hand as she fell, her ballet shoe skidding in the spilled water. She only cried out once, with surprise more than pain, but she whimpered and held her arms over her head as he loomed over her, his arm raised again.

She went through her mental routine. Don't cry, don't fight. You fight, he'll hurt you. Don't cry, don't scream, neighbors'll know, people will hear, people will know. Guilty and ashamed and protective of her body, she pressed her lips tight together, so that no one would know that Roy Wattles was not only a loud town drunk. No one would know what life was like in that bungalow in Armenia, Georgia, for Roy Wattles and his daughter, Mary Louise. So she crouched and cringed and willed herself to faint so she couldn't feel him pick her up and drag her across the floor.

"A-*round* and a-*round* and a-*round* and a-*round*.... Good! Feel those neck muscles relax now? What about you, Bonnie? Y'all feeling those muscles loosen now?"

"Yes, Miss Mary Louise," they chorused.

"Then *o*-ver and *down* and *o*-ver and *down*, a-*gain*—" She glanced around as she bent with them. Patty-Warren was bent over but she wasn't bobbing like the others. She seemed frozen, like one of those rubber dolls that can be bent into any position. Mary Louise was about to say something but stopped herself. The others might tease the girl. Maybe she is lazy, Mary Louise thought. If she were sick, her mother surely wouldn't have let her come. She went on with the class.

When it was over, Patty-Warren was one of the last of the children to retrieve her dance bag and leave. Mary Louise noticed that Patty-Warren's bag was of soft chamois, while all the others were made of nylon or canvas.

"You seemed kinda stiff today, Patty-Warren," Mary Louise said.

Patty-Warren looked up and Mary Louise caught her breath. It was such a strange look, such a haunted one. Not at all the look of a child of six. Still she didn't speak.

Mary Louise went to her and knelt down on one knee.

"Patty-Warren, you like to dance?" she asked softly.

And then the face changed. Patty-Warren was suddenly a delighted, beautiful little girl. "I love to," she said softly.

One of the children had stopped in the doorway and stood unabashedly listening to the exchange. Now that there was whispering, she moved in closer.

Mary Louise turned. "Go on home, Natalie," she said. "You go on, now. Your mama's waiting." Only when the child had reluctantly plodded out did Mary Louise turn again to Patty-Warren, who was neatly placing her folded legwarmers in her dance bag.

"If you love to dance, then you're starting out at the best age," Mary Louise continued as the child patted her belongings in the bag. "That's what Miss Dorothy says. But you have to work hard on your—"

"Patty-Warren Stokely! My goodness, I've been standing out there in the hall forever!" The voice came from the doorway and made both girls jump. "Seems every other child is long gone!"

"I'm coming, Mama," Patty-Warren answered, and brushing quickly past Mary Louise, she headed toward the door, swinging her dance bag over her shoulder.

When Mary Louise arrived for her own Monday class, she heard angry voices coming from the office and she recognized one of them.

"...can't imagine what you're talking about!" (Miss Dorothy)

"I'll just bet you can't! I just want to hear what some of the other mothers have to say!" (Someone Else)

"Now, please, all you have to do is watch—" (Miss Dorothy)

"I'm going to talk to my husband about taking you to court!" The door opened and the Someone Else, who turned out to be Mrs. Stokely, Patty-Warren's mother, stormed out, leaving the door open.

Mary Louise went in.

"What on earth was that all about?" she asked.

Miss Dorothy leaned heavily against a big oak desk, the only thing in the room that belonged to the churches. "That woman's got to be crazy," she said, shaking her head back and forth. "She barged in here and accused us of 'working' the children too hard. Said it was practically forced labor! Can you believe it? She said her daughter, Patty, was just exhausted, achy, bruised—all but fainting Friday night."

"Patty-Warren barely does the warm-ups," Mary Louise said. "Honestly, I can't get her to move!"

"Well, what is the matter with that woman? Did you hear her mention *court?* Really, Mary Louise, you know the Stokelys have only been here a few months, but they do carry weight in town. Everyone knows who they are. If she spreads word that we're hurting the children— Oh, she makes me so mad! I tried to get her to watch a class, but you saw! She just stormed out!"

"I saw," Mary Louise said. She was picturing the two faces of Patty-Warren Stokely: one, odd, haunted the other...*Do you like to dance, Patty-Warren? I love to.*

"Well, I'll tell you something, Mary Louise," Miss Dorothy was saying. "She won't hurt us. Believe me. There isn't a thing she can do, because she's just plain wrong! And I'll fight her and anybody else for what we've got going here—because it's right!"

Mary Louise glowed. Miss Dorothy had said "we," and Mary Louise knew she was being included. She studied the dance teacher. No one would ever guess her age, whatever it was, with her lean dancer's body, that chestnut hair pulled tight into a ponytail. She would fight, she'd fight anybody—all by herself if she had to—and here was some-one you'd want to have on your side.

"Yes, *ma'am,*" Mary Louise said with a nod.

On Friday, Patty-Warren was absent from jazz class and so were Marcia Willis, Sally Munro, and Natalie Laroquette. Miss Dorothy looked in on the class briefly, shook her head, rolled her eyes, and ducked out again.

Mary Louise, on the floor bicycling with the remaining students, thought: I have to see that child again. I have to see that child again.

There isn't much mystery to most six-year-olds, Mary Louise thought. They haven't lived that long. Yet here was one. A pretty mama with pretty clothes, a rich daddy, a big house—that old Bishop place they did over—yet she doesn't seem spoiled rotten. She doesn't run the group—even gets teased by them. She doesn't do warm-up exercises, but her mama says she's worked like a donkey. Doesn't fig-ure up. And those two faces. Especially the first one, that odd one—where have I seen it before?...I have to see that child again.

After class, Mary Louise hurriedly pulled on her jeans over her tights and whipped off the sweatband from under her curly bangs.

"Where are you off to in such a hurry?" Miss Dorothy asked.

"Nowhere special," Mary Louise grunted.

"Uh-huh, and you're just fine, right?"

"Right."

"And everything's fine at home, right?"

"That's right."

"And you aren't going anywhere special."

"Right again."

"Mary Louise, I told you I don't pry..."

"Yes, ma'am."

"But I told you that when you need help you've got it, didn't I?"

"Yes, ma'am."

"Just remember that."

Mary Louise looked up at Miss Dorothy and caught a glimpse of herself in the big mirror on the far wall of the dance room. It was her own face, her own, still wearing the expression it had shown to Miss Dorothy, but suddenly Mary Louise couldn't breathe. She recognized the face, and the pain she felt was actually physical. She knelt quickly and pretended to fiddle with her shoe to hide what she was feeling from Miss Dorothy.

"Here come the girls," Miss Dorothy said, turning away. "See you Monday, Mary Louise."

"I don't want her to see you," Mrs. Stokely said. "She's in bed. Poor little thing—still in pain from last week's dance class. You people are going to regret this, I promise you. My husband and I are going to take this further!"

"Please, Mrs. Stokely," Mary Louise begged. She stood shivering in the doorway of the old Bishop house, even though the evening was warm and she had tights on under her jeans. "All's I want is to talk to Patty-Warren, just for a few minutes. Just cheer her up a little if she's feeling bad. I got to care for her, Mrs. Stokely, even though I haven't known her that long. Please, Mrs. Stokely, *please*!" It didn't matter to her if Mrs. Stokely refused her. Mary Louise would see that child if it meant climbing up the vines on the side of the house in the middle of the night. She would see her if it meant hiding in the bushes till winter! If it meant waiting on the grounds of this house forever and ever, Mary Louise knew she couldn't go anywhere, do anything, until she had some time alone with Patty-Warren Stokely, aged six.

But Mrs. Stokely was tired. She wasn't in the mood for fighting with this small-town teenager. She wasn't much in the mood for anything. She sighed, backed away, and let Mary Louise in.

Patty-Warren's room was the size of Mary Louise's whole bungalow. It was all pink, green, and white and filled with lace and eyelet and chintz and ruffles and stuffed animals and dolls.

In the middle of a pink, green, and white quilted canopied bed, Patty-Warren sat, hunched over a clown doll. She glanced up, and Mary Louise touched her chest with her fingers. On the child's face was the look she had seen on her own face in the mirror not fifteen minutes ago. The same look.

A sixteen-year-old, a six-year-old. One dirt poor, the other rich by all standards. One with an absentee mother and a crazed drunk for a father, the other with doting parents. But their faces—their eyes— were the same. And Mary Louise knew it at once.

"Hello, there, Patty-Warren," she said, moving slowly toward the child on the bed. "Honey, this is just about the prettiest room I ever did see!" Patty-Warren only stared at her. "Could I sit here? On your bed with you? Only if you say it's okay, though." Mary Louise turned her head. There was no one in the doorway.

"Could I sit, Patty-Warren? *May I*? Isn't that what they're always tellin' us? Say *may* I, not *can* I or *could* I? Doesn't your mama say that?" She glanced again toward the door.

Patty-Warren nodded. The corners of her mouth were turned down. She glanced toward the edge of her bed and Mary Louise understood without words that it was all right if she sat down.

"That's a cute doll, Patty-Warren," Mary Louise said, and smiled at the clown in the child's lap. She felt as though her heart would pound right through her chest, but she was more than adept at hiding fear, and outwardly she was calm and quiet. Just like the child.

"Patty-Warren, your mama says you're home here because you hurt yourself in dance class last week."

The child clutched the doll tightly and looked at her knees.

Mary Louise dropped her voice to almost a whisper. "But it doesn't hurt you to dance, does it? You love to dance, don't you, Patty-Warren?"

"Yessss," the child breathed.

"I know it," Mary Louise said. "I know it. Me, too. I used to dance all by myself before Miss Dorothy came to town. It was my own secret. Used to do it out in the back lot near our house, where the trees and bushgrowth hid me from the road. Never loved anything more—you know what I mean, Patty-Warren?"

The child only looked at her, but Mary Louise knew she knew.

"Used to think I could make my body float right away from the ground. Made my body move like a swaying flower, like a thing without any bones at all. But...the thing of it is..." She picked up the scalloped edge of Patty-Warren's pink sheet and began to twirl it with her fingers. "The thing of it is, I found out if I fight my daddy, then he hurts me. Hurts my body. And if he does that, Patty-Warren...if he hurts me...then...I can't dance. Hurts to bend, you know? Hurts to move..."

It seemed to Mary Louise that Patty-Warren wasn't breathing.

"So anyway," she went on, "what I learned to do, I learned not to fight." She said it with a shrug, in a matter-of-fact voice. "And then, no matter how much I hate what my daddy's doin', well...I still get to dance. If I don't fight."

Patty-Warren hadn't moved except for her fingers, which were now grasping the arm of her toy clown.

"My mama never knew about it, either, just like yours," Mary Louise continued. "'Course it was different with me. My mama just left 'cause my daddy drank and beat on her, not 'cause of what was goin' on with me. He told me never to tell. Just like your daddy told you, right?"

Patty-Warren swallowed.

"It hurts more when you're little, I know," Mary Louise said. "So you fight and that makes it worse." She took the little girl's chin and tilted it upward. "You never told on your daddy, did you, Patty-Warren? Just like he told you. You were a good girl, just like me. Neither one of us ever told, we were good girls, Patty-Warren, isn't that right."

Without warning, Patty-Warren broke, flung her body, skinny little arms outstretched, and hugged Mary Louise's neck for all she was worth, weeping hot tears into the collar of the older girl's shirt.

After a while Mary Louise took the little face in her two hands. "Patty-Warren, we're still good girls, you and me. Here's how we can do it so's it'll be all right: I'll be the one to tell for you. I'll do it. And the only thing you have to do is say, 'Yes, ma'am, that's right,' when I'm done. That's not telling—*I'll* do the telling. You see?"

Mary Louise walked out of the big house, walked slowly back toward town, toward the lights of the Armenia Town Center where they were getting ready for a jukebox party for the junior high...where the last of the tap and jazz and ballet classes would have just finished...and where Miss Dorothy would be sprawled on the green leather couch in the office, sipping a Coke, wiggling her toes, listening to a record. And she'd be almost but not quite ready to head for home.

Mary Louise still felt Patty-Warren's arms around her neck, and her collar was damp from the child's tears. But what she remembered—what she would always remember—was the child's answer as they hugged each other: "And then...Miss Mary Louise...do I tell for you?"

Alan Syliboy *All My Relations, Family,* 1992

And So I Go

BY

PATRICIA

GRACE

ur son, brother, grandchild, you say you are going away from this place you love, where you are loved. Don't go. We warm you. We give you strength, we give you love.

These people are yours.

These hills, this soil, this wide stretch of sea.

This quiet place.

This land is mine, this sea, these people. Here I give love and am loved but I must go, this is in me. I go to learn new ways and to make a way for those who follow because I love.

My elders, brothers and sisters, children of this place, we must go on. This place we love cannot hold us always. The world is large. Not forever can we stay here warm and quiet to turn the soil and reap the sea and live our lives. This I have always known. And so I go ahead for those who come. To stand mid-stream and hold a hand to either side. It is in me. Am I not at once dark and fair, fair and dark? A mingling. Since our blue-eyed father held our dark-eyed mother's hand and let her lead him here.

But, our brother, he came, and now his ways are hers out of choice because of love.

And I go because of love. For our mother and her people and for our father. For you and for our children whose mingling will be greater than our own. I make a way. Learn new ways. So I can't take up that which is our father's and hold it to the light. Then the people of our mother may come to me and say, "How is this?" And I will hold the new thing to the light for them to see. Then take up that which is our mother's and say to those of our father, "You see? See there, that is why."

And brother, what of us. Must we do this too? Must we leave this quiet place at the edge of hills, at the edge of sea and follow you? For the sake of our mother's people who are our own? And for our father and because we love?

You must choose but if you do not feel it in you, stay here in warmth. Let me do this and do not weep for my going. I have this power in me. I am full. I ache for this.

Often I have climbed these hills and run about as free as rain. Stood on the highest place and looked down on great long waves looping on to sand. Where we played, grew strong, learned our body skills. And learned the ways of summers, storms, and tides. From where we stepped into the spreading sea to bathe or gather food. I have watched and felt this ache in me.

I have watched the people. Seen myself there with them living too. Our mother and our blue-eyed father who came here to this gentle place that gives us life and strength. Watched them work and play, laugh and cry, and love.

Seen our uncle sleeping. Brother of our mother. Under a tree bright and heavy with sunned fruit. And beside our uncle his newest baby daughter sleeping too. And his body-sweat ran down and over her head in a new baptising. I was filled with strength.

And old Granny Roka sits on her step combing her granddaughter's hair, patiently grooming. Plaiting and tying the heavy tangled kelp which is her pride. Or walk together on the mark of tide, old Granny and the child, collecting sun-white sticks for the fire. Tying the sticks into bundles and carrying them on their backs to the little house. Together.

And seen the women walk out over rocks when the tide is low, submerging by a hole of rock with clothes ballooning. Surfacing with wine-red crayfish, snapping tails and clawing air on a still day. And on a special day the river stones fired for cooking by our father, our cousins, and our uncles who laugh and sing. Working all as one.

Our little brother's horse walks home with our little one asleep. Resting a head on his pony's neck, breathing in the warm horse stink, knees locked into its sides. Fast asleep on the tired flesh of horse. And I ache. But not forever this. And so I go.

And when you go our brother as you say you must will you be warm? Will you know love? Will an old woman kiss your face and cry warm tears because of who you are? Will children take your hands and say your name? In your new life our brother will you sing?

The warmth and love I take from here with me and return for their renewal when I can. It is not a place of loving where I go, not the same as love that we have known.

No love fire there to warm one's self beside
No love warmth
Blood warmth
Wood and tree warmth
Skin on skin warmth
Tear warmth
Rain warmth
Earth warmth
Breath warmth
Child warmth
Warmth of sunned stones
Warmth of sunned water
Sunned sand
Sand ripple
Water ripple
Ripple sky
(Sky Earth
Earth Sky
And our beginning)

And you ask me shall I sing. I tell you this. The singing will be here within myself. Inside this body. Fluting through these bones. Ringing in the skies of being. Ribboning in the course of blood to soothe swelled limbs and ache bruised heart.

You say to us our brother you will sing. But will the songs within be songs of joy? Will they ring? Out in the skies of being as you say? Pipe through bone, caress flesh wounding? Or will the songs within be ones of sorrow.

Of warmth dreams
Love dreams
Of aching
And flesh bruising.
If you listen will it be weeping that you hear?
Lament of people
Earth moan
Water sigh
Morepork cry of death?

My sisters, brothers, loved ones, I cannot tell. But there will be gladness for me in what I do. I ask no more. Some songs will be of joy and others hold the moan and sigh, the owl cry and throb of loneliness.

What will you do then our brother when the singing dirges through your veins, pressing and swelling in your throat and breast, pricking at your mind with its aching needles of sound?

What should I do but deny its needling and stealing into mind. Its pressing into throat and breast. I will not put a hand of comfort over body hardenings nor finger blistered veins in soothing. The wail, the lament shall not have my ear. I will pay the lonely body ache no mind. Thus I go.

I stand before my dark-eyed mother, blue-eyed father, brothers, sisters. My aunts and uncles and their children and these old ones. All the dark-eyed, light-eyed minglings of this place.

We gather. We sing and dance together for my going. We laugh and cry. We touch. We mingle tears as blood.

I give you my farewell.

Now I stand on a tide-wet rock to farewell you sea. I listen and hear your great heart thud. I hear you cry. Do you too weep for me? Do you reach out with mottled hands to touch my brow and anoint my tear-wet face with tears of salt? Do not weep but keep them well. Your great heart beats I know for such as these. Give them sea, your great sea love. Hold them gently. Already they are baptised in your name.

As am I
And take your renewal where I go
And your love

Take your strength
And deep heart thud
Your salt kiss
Your caring.

Now on a crest of hill in sweeping wind. Where I have climbed and run. And loved and walked about. With life brimming full in me as though I could die of living.

Guardian hill you do not clutch my hand, you do not weep. You know that I must go and give me blessing. You guard with love this quiet place rocking at the edge of sea...

And now at the highest place I stand. And feel a power grip me. And a lung-bursting strength. A trembling in my legs and arms. A heavy ache weighting down my groin.

And I lie on soil in all my heaviness and trembling. Stretch out my arms on wide Earth Mother and lay my face on hers. Then call out my love and speak my vow.

And feel release in giving to you earth, and to you sea, to these people.

So I go. And behind me the sea-moan and earth-cry, the sweet lament of people. Towards the goddess as she sleeps I go. On with light upon my face.

The Last Song

BY

ELTON JOHN AND

BERNIE TAUPIN

Yesterday you came to lift me up
As light as straw and brittle as a bird
Today I weigh less than a shadow on the wall
Just one more whisper of a voice unheard

Tomorrow leave the windows open
As fear grows please hold me in your arms
Won't you help me if you can to shake this anger
I need your gentle hands to keep me calm

'Cause I never thought I'd lose
I only thought I'd win
I never dreamed I'd feel
This fire beneath my skin
I can't believe you love me
I never thought you'd come
I guess I misjudged love
Between a father and his son

Things we never said come together
The hidden truth no longer haunting me
Tonight we touched on the things that were
never spoken
That kind of understanding sets me free

'Cause I never thought I'd lose
I only thought I'd win
I never dreamed I'd feel
This fire beneath my skin
I can't believe you love me
I never thought you'd come
I guess I misjudged love
Between a father and his son

A Candle in a Glass

BY

MARGE

PIERCY

When you died, it was time to light the first
candle of the eight. The dark tidal shifts
of the Jewish calendar of waters and the moon
that grows like a belly and starves like a rabbit
in winter have carried that holiday forward
and back since then. I light only your candle
at sunset, as the red wax of the sun melts
into the rumpled waters of the bay.

The ancient words pass like cold water
out of stone over my tongue as I say kaddish.
When I am silent and the twilight drifts
in on skeins of unraveling woolly snow
blowing over the hill dark with pitch pines,
I have a moment of missing that pierces
my brain like sugar stabbing a cavity
till the nerve lights its burning wire.

Grandmother Hannah comes to me at Pesach
and when I am lighting the sabbath candles.
The sweet wine in the cup has her breath.
The challah is braided like her long, long hair.
She smiles vaguely, nods, is gone like a savor
passing. You come oftener when I am putting
up pears or tomatoes, baking apple cake.
You are in my throat laughing or in my eyes.

When someone dies, it is the unspoken words
that spoil in the mind and ferment to wine
and to vinegar. I obey you still, going
out in the saw toothed wind to feed the birds
you protected. When I lie in the arms of my love,
I know how you climbed like a peavine twining,
lush, grasping for the sun, toward love
and always you were pinched back, denied.

It's a little low light the yahrzeit candle
makes, you couldn't read by it or even warm
your hands. So the dead are with us only
as the scent of fresh coffee, of cinnamon,
of pansies excites the nose and then fades,
with us as the small candle burns in its glass.
We lose and we go on losing as long as we live,
a little winter no spring can melt.

Some Long Journey

BY

JENNY
BORNHOLDT

There was the time
the family floated down
the river.
We were all there then—
my parents, my sisters
and I. We lay on our backs
and the broad water
carried us
down towards the sea.
It was summer
and the day was
spacious, with room
for everyone.

Even my two
grandfathers were there
and my grandmothers,
enjoying the roll
of the water.
They turned in to the
bank early and lay
resting on the warm
stones. We went on
enjoying the pull
of the water, its
casual insistence
all of us relaxed,
laughing, even, at our
willingness to be
carried that way.

We rode all the way
to the river mouth
where we struck ground
the sea coming to meet
us over the stones.
And we felt satisfied at
having come that
distance together,
lay for a while, the
sun warming our bodies
through the moving water.

ACKNOWLEDGEMENTS

Care has been taken to trace ownership of copyright material contained in this text. The publishers will gladly accept any information that will enable them to rectify any reference or credit in subsequent editions.

TEXT

p. 1 "Offered and Taken" by Dominique LaBaw. Dominique LaBaw was named one of the best new poets of the 1990s by *LA Weekly* magazine. She lives in Los Angeles; **p. 2** "What Happened to Family?" by Lesley Francis. Reprinted with permission from The Edmonton Journal; **p. 4** "Friends and Family" by Susan Rogers, *Canadian Living*, 1994; **p. 10** "Mothers" by Nikki Giovanni from *My House* by Nikki Giovanni. Copyright © 1972 by Nikki Giovanni. By permission of William Morrow and Company, Inc.; **p. 12** "Whose Mouth Do I Speak With" by Suzanne Rancourt. From *Callaloo*, Vol. 17, No. 1, 1994. Reprinted by permission of The John Hopkins University Press; **p. 14** Lucille Clifton. Excerpt from "generations," copyright © 1987, by Lucille Clifton. Reprinted from *Good Woman: Poems and a Memoir 1969-1980*, by Lucille Clifton, with the permission of BOA Editions, Ltd., 92 Park Ave., Brockport NY 14420; **p. 20** "Father's Gloves" by Paul Wilson. Reprinted from *Dreaming My Father's Body* (Coteau Books, 1994) by Paul Wilson, with permission of the publishers; **p. 22** "The Boat" by Alistair MacLeod. From *The Lost Salt Gift of Blood* by Alistair MacLeod. Used by permission of the Canadian Publishers, McClelland & Stewart, Toronto; **p. 40** "My Collection" by Dinyar Godrej. Reprinted with permission from New Internationalist, April 1992; **p. 42** "Faces" by Julia Lowry Russell. First appeared in *Earth Song, Sky Spirit*. Reprinted by permission of Julia Lowry Russell; **p. 47** "Where My Father Sat" by Dave Margoshes from *Walking at Brighton* (Thistledown Press Ltd., 1988), used with permission; **p. 48** "Grief and Fear" by Michael Coren, *Globe and Mail* columnist and author; **p. 51** "Learning to Drive" by Mark Vinz. First published in *Beyond Borders: An Anthology of New Writing from Manitoba, Minnesota, Saskatchewan and the Dakotas* edited by Mark Vinz and Dave Williamson (Turnstone Press, Winnipeg, and New Rivers Press, Minneapolis, 1992); **p. 52** "Now, A Word From Mother" by Megan Rosenfeld © 1993 The *Washington Post*. Reprinted with permission; **p. 56** "A Sweet Sad Turning of the Tide" by Maude Meehan. From *The Tie That Binds: A Collection of Writings about Fathers and Daughters, Mothers and Sons*, edited by Sandra Martz. Reprinted by permission of Papier Mache Press; **p. 58** "Standard Answers" by Peg Kehret. From *Acting Natural* by Peg Kehret. Copyright © 1991 Meriwether Publishing Ltd., Colorado Springs, Colorado 80907; **p. 62** "April 19th, 1985" by Deirdre Levinson. Reprinted by permission of the author; **p. 66** "January Chance" from *Collected Poems 1924-1963* by Mark Van Doren. Copyright © 1963 by Mark Van Doren and copyright renewed © 1991 by Dorothy G. Van Doren. Reprinted by permission of Hill and Wang, a division of Farrar, Straus & Giroux, Inc.; **p. 67** "Pockets" by Karen Swenson. Copyright 1989 Karen Swenson; **p. 68** "Our Eldest Son Calls Home" by Don Polson from *Moving Through Deep Snow* (Thistledown Press Ltd., 1984), used with permission; **p. 70** "I'm Not Slave Material" by Eileen Gilchrist. By permission of Eileen Gilchrist; **p. 73** "Dear Mom: This Is the Letter I Would Have Written" by Monica Steel. Monica Steel is a pseudonym for a Toronto writer; **p. 76** "Ben: A Monologue" by Roger Karshner. From *Teenage Mouth* © 1993 by Dramaline Publications. Reprinted by permission of Dramaline Publications, 36-851 Palm View Road, Rancho Mirage, CA

92270; **p. 78** "Strangers" by Michael Ngo. From *New Canadian Voices*, edited by Jessie Porter. Reprinted with the permission of the publisher, Wall & Emerson, Inc., Toronto; **p. 82** "A Villager's Response" by Steve Owad. Originally published in the *University of Windsor Review*, Vol. 25, No. 1&2. Reprinted with permission; **p. 89** "A Moving Day" by Susan Nunes from *The Graywolf Annual Seven: Stories from the American Mosaic* edited by Scott Walker, 1990. Reprinted by permission of the author; **p. 96** "My Grandparents" by Ruth Fainlight. From *Selected Poems* by Ruth Fainlight. Reprinted by permission of the publisher, Hutchinson; **p. 97** "A Poet's Story" by Julio Henriquez. From *Refugees*, No. 79/October 1990, UNHCR; **p. 101** "Sisters" by Sibani Raychaudhuri. From *Flaming Spirit* by the Asian Women Writers' Collective, first published by Virago, 1994, The Rotunda, 42-43 Gloucester Crescent, Camden Town, London NW17PD; **p. 113** "Sister" by Carol Shields. From *Intersect* by Carol Shields, 1974. Reprinted by permission of Borealis Press Ltd., Canada; **p. 114** "The Secret World of Siblings" by Erica E. Goode. Copyright, January 10, 1994, U.S. News & World Report; **p. 124** "Brothers" by Bret Lott. Copyright © 1993 by the Antioch Review, Inc. First appeared in the *Antioch Review*, Vol. 51, No. 1 (Winter, 1993). Reprinted by permission of the Editors; **p. 131** "My Sister's Hand" by Jill Rinehart. Reprinted by permission of Jill Rinehart; **p. 138** "To My Dear and Loving Husband" by Anne Bradstreet. Reprinted by permission of the publishers from *The Works of Anne Bradstreet* edited by Jeannine Hensley, Cambridge, Mass.: The Belknap Press of Harvard University Press, Copyright © 1967 by the President and Fellows of Harvard College; **p. 139** "Spaghetti" by Nancy Botkin. Originally published in the *University of Windsor Review*, Vol. 25, No. 1&2. Reprinted with permission; **p. 140** "Heirloom" by Eamon Grennan. Reprinted by permission; © 1993 Eamon Grennan. Originally in *The New Yorker*; **p. 142** "Fathers" by Lake Sagaris. From *Celebrating Canadian Women* edited by Greta Hofmann Nemiroff, 1989. Reprinted by permission of Fitzhenry & Whiteside. **p. 144** "His, Hers & Theirs" by Leslie Blake-Côté, Writer/Consultant/Teacher; **p. 147** "Ties" by Shirley A. Serviss is reprinted from *Model Families* by permission of Rowan Books—an imprint of The Books Collective; **p. 148** "'How We Adopted Me'" by Jane Marks. Copyright © 1993 by The New York Times Company. Reprinted by permission. Reprinted by permission of Jane Marks, author of *The Hidden Children: The Secret Survivors of the Holocaust* (Ballantine Books); **p. 154** "What Feels Like the World" by Richard Bausch. Copyright © 1987 by Richard Bausch. Reprinted by permission of Simon & Schuster, Inc.; **p. 169** "Woman With No Face" by Alice Lee. First published in *Gatherings*, Vol. III, 1992. Reprinted by permission of the author; **p. 172** "Hero" by Sheree Fitch is reprinted from *In This House Are Many Women* by permission of Goose Lane Editions copyright © Sheree Fitch, 1993; **p. 176** "Power Failure" by Jane Rule from *Inland Passage*, Naiad Press, 1985, is reprinted with the permission of the author and publisher; **p. 187** "They're Mothering Me to Death" by Eileen Herbert Jordan. By permission of Eileen Herbert Jordan; **p. 190** "Sandwiched" by Suzan Milburn. By permission of the author. Suzan Milburn lives in Vernon, B.C., and has been writing poetry for eight years; **p. 192** "The Day I Married..." by Jo Carson copyright © 1989 by Jo Carson, from *Stories I Ain't Told Nobody Yet* by Jo Carson. Reprinted by permission of Orchard Books, New York; **p. 194** *Mama Makes Up Her Mind: And Other Dangers of Southern Living* (pp. 68-78), © 1993 by Bailey White. Reprinted by permission of Addison-Wesley Publishing Company, Inc.; **p. 200** "Daddytrack" by John Byrne Barry. Originally appeared in *Mothering* Magazine. Reprinted by permission of John Byrne Barry; **p. 204** "His Kitchen" by Annharte.

Reprinted with permission from *Being on the Moon* published by Polestar Press Ltd., Vancouver, B.C.; **p. 206** "How Poverty Affects My Kids" by Anonymous. Reprinted with permission from *No Way To Live: Poor Women Speak Out* by Sheila Baxter, published by New Star Books Ltd., Vancouver, B.C., 1988; **p. 210** "The Kids Are All Right" by Susan Faludi. Reprinted with permission from *Mother Jones* magazine, © 1988, Foundation for National Progress; **p. 214** "Kids Have Rights Too!" by Marcia Kaye. Reprinted by permission of Marcia Kaye; **p. 218** "The Child Who Walks Backwards" by Lorna Crozier. From *The Garden Going On Without Us* by Lorna Crozier. Used by permission of the Canadian Publishers, McClelland & Stewart, Toronto; **p. 220** "The Good Girls" by Fran Arrick, copyright © 1987 by Fran Arrick. From *Visions* by Donald R. Gallo, Editor. Used by permission of Dell Books, a division of Bantam Doubleday Dell Publishing Group, Inc.; **p. 233** "And So I Go" from *Waiariki* by Patricia Grace, 1975. Reproduced with permission from Longman Paul Ltd.; **p. 238** "THE LAST SONG" (Elton John and Bernie Taupin) © 1992 HAPPENSTANCE LIMITED & HANIA. All rights administered by WARNER/CHAPPELL MUSIC CANADA LIMITED. All rights reserved, Used By Permission; **p. 240** "A Candle in a Glass" by Marge Piercy. From *Available Light* by Marge Piercy. Copyright © 1988 by Middlemarsh Inc. Reprinted by permission of Alfred A. Knopf Inc.; **p. 242** "Some Long Journey" by Jenny Bornholdt. Copyright Jenny Bornholdt, 1992.

PHOTOGRAPHS
p. 5 Mark Mainguy; **p. 13** Greg Staats; **p. 19** Ontario Black History Society/Daniel Hill; **p. 26** 51.1 Donald Cameron MacKay *Landscape, Herring Cove* c 1950 oil on canvas 61.1 x 76.1 cm. Collection of The Art Gallery of Nova Scotia; **p. 50** Dick Hemingway; **p. 61** FOR BETTER OR FOR WORSE © Lynn Johnston Prod., Inc. Reprinted with permission of UNIVERSAL PRESS SYNDICATE. All rights reserved; **p. 81 & p. 86** Robert Garrard; **p. 100** CIDA Photo/David Barbour; **p. 112** Dick Hemingway; **p. 123** Robert Garrard; **p. 130** Dick Hemingway; **p. 137** Shelly Niro/Native Indian/Inuit Photographers' Association *Portrait #4: Family: Stylistic Beadwork Designs and Blue Lights*; **p. 171** Dorothy Chocolate/Communications Society of the Western NWT/Native Indian/Inuit Photographers' Association; **p. 175** Dick Hemingway; **p. 203** CALVIN AND HOBBES © Watterson. Reprinted with permission of UNIVERSAL PRESS SYNDICATE. All rights reserved; **p. 205** © Jean Mahaux/The Image Bank Canada; **p. 217** Robert Garrard; **p. 232** Alan Syliboy, Millbrook, Nova Scotia, 1952 *All My Relations, Family*, 1992, 4 serigraph prints each 41.0 x 51.0 cm. Collection of the artist.